Shimon Whiteson

Adaptive Representations for Reinforcement Learning

Studies in Computational Intelligence, Volume 291

Editor-in-Chief
Prof. Janusz Kacprzyk
Systems Research Institute
Polish Academy of Sciences
ul. Newelska 6
01-447 Warsaw
Poland
E-mail: kacprzyk@ibspan.waw.pl

Further volumes of this series can be found on our homepage: springer.com

Vol. 270. Zong Woo Geem
Recent Advances In Harmony Search Algorithm, 2009
ISBN 978-3-642-04316-1

Vol. 271. Janusz Kacprzyk, Frederick E. Petry, and Adnan Yazici (Eds.)
Uncertainty Approaches for Spatial Data Modeling and Processing, 2009
ISBN 978-3-642-10662-0

Vol. 272. Carlos A. Coello Coello, Clarisse Dhaenens, and Laetitia Jourdan (Eds.)
Advances in Multi-Objective Nature Inspired Computing, 2009
ISBN 978-3-642-11217-1

Vol. 273. Fatos Xhafa, Santi Caballé, Ajith Abraham, Thanasis Daradoumis, and Angel Alejandro Juan Perez (Eds.)
Computational Intelligence for Technology Enhanced Learning, 2010
ISBN 978-3-642-11223-2

Vol. 274. Zbigniew W. Raś and Alicja Wieczorkowska (Eds.)
Advances in Music Information Retrieval, 2010
ISBN 978-3-642-11673-5

Vol. 275. Dilip Kumar Pratihar and Lakhmi C. Jain (Eds.)
Intelligent Autonomous Systems, 2010
ISBN 978-3-642-11675-9

Vol. 276. Jacek Mańdziuk
Knowledge-Free and Learning-Based Methods in Intelligent Game Playing, 2010
ISBN 978-3-642-11677-3

Vol. 277. Filippo Spagnolo and Benedetto Di Paola (Eds.)
European and Chinese Cognitive Styles and their Impact on Teaching Mathematics, 2010
ISBN 978-3-642-11679-7

Vol. 278. Radomir S. Stankovic and Jaakko Astola
From Boolean Logic to Switching Circuits and Automata, 2010
ISBN 978-3-642-11681-0

Vol. 279. Manolis Wallace, Ioannis E. Anagnostopoulos, Phivos Mylonas, and Maria Bielikova (Eds.)
Semantics in Adaptive and Personalized Services, 2010
ISBN 978-3-642-11683-4

Vol. 280. Chang Wen Chen, Zhu Li, and Shiguo Lian (Eds.)
Intelligent Multimedia Communication: Techniques and Applications, 2010
ISBN 978-3-642-11685-8

Vol. 281. Robert Babuska and Frans C.A. Groen (Eds.)
Interactive Collaborative Information Systems, 2010
ISBN 978-3-642-11687-2

Vol. 282. Husrev Taha Sencar, Sergio Velastin, Nikolaos Nikolaidis, and Shiguo Lian (Eds.)
Intelligent Multimedia Analysis for Security Applications, 2010
ISBN 978-3-642-11754-1

Vol. 283. Ngoc Thanh Nguyen, Radoslaw Katarzyniak, and Shyi-Ming Chen (Eds.)
Advances in Intelligent Information and Database Systems, 2010
ISBN 978-3-642-12089-3

Vol. 284. Juan R. González, David Alejandro Pelta, Carlos Cruz, Germán Terrazas, and Natalio Krasnogor (Eds.)
Nature Inspired Cooperative Strategies for Optimization (NICSO 2010), 2010
ISBN 978-3-642-12537-9

Vol. 285. Roberto Cipolla, Sebastiano Battiato, and Giovanni Maria Farinella (Eds.)
Computer Vision, 2010
ISBN 978-3-642-12847-9

Vol. 286. Zeev Volkovich, Alexander Bolshoy, Valery Kirzhner, and Zeev Barzily
Genome Clustering, 2010
ISBN 978-3-642-12951-3

Vol. 287. Dan Schonfeld, Caifeng Shan, Dacheng Tao, and Liang Wang (Eds.)
Video Search and Mining, 2010
ISBN 978-3-642-12899-8

Vol. 288. I-Hsien Ting, Hui-Ju Wu, Tien-Hwa Ho (Eds.)
Mining and Analyzing Social Networks, 2010
ISBN 978-3-642-13421-0

Vol. 289. Anne Håkansson, Ronald Hartung, and Ngoc Thanh Nguyen (Eds.)
Agent and Multi-agent Technology for Internet and Enterprise Systems, 2010
ISBN 978-3-642-13525-5

Vol. 290. Weiliang Xu and John Bronlund
Mastication Robots, 2010
ISBN 978-3-540-93902-3

Vol. 291. Shimon Whiteson
Adaptive Representations for Reinforcement Learning, 2010
ISBN 978-3-642-13931-4

Shimon Whiteson

Adaptive Representations for Reinforcement Learning

Springer

Dr. Shimon Whiteson
Universiteit van Amsterdam
Instituut voor Informatica
Science Park 107
1098 XG Amsterdam
Netherlands
E-mail: s.a.whiteson@uva.nl

ISBN 978-3-642-42231-7 ISBN 978-3-642-13932-1 (eBook)

DOI 10.1007/978-3-642-13932-1

Studies in Computational Intelligence ISSN 1860-949X

Softcover re-print of the Hardcover 1st edition 2010

Typeset & Cover Design: Scientific Publishing Services Pvt. Ltd., Chennai, India.

Printed on acid-free paper

9 8 7 6 5 4 3 2 1

springer.com

To Adam and Rena

Foreword

It is a great pleasure and honor to write the forward for this book, representing the culmination of Shimon Whiteson's Ph.D. thesis research at The University of Texas at Austin. When I arrived at UT Austin in the fall for 2002, Shimon was one of the first students to walk into my office and we began doing research together almost immediately. Our research helped form the nucleus of my research group, the Learning Agents Research Group, and Shimon became my first Ph.D. graduate in the spring of 2007.

Shimon was an ideal first student for a new assistant professor. He has a strong sense of what he wants to learn, and is never satisfied with a partial answer to any question. Most importantly for this book, he is self-assured and is willing to take risks in order to achieve meaningful results.

After several initial contributions that built upon my own past research, Shimon set off on his own towards his most important technical contribution so far, namely the development of a novel algorithm, NEAT+Q, capable of evolving neural network function approximators for reinforcement learning agents. The technical details of NEAT+Q are well-presented in the book, so suffice it to say here that his work on adaptive representations for reinforcement learning takes a substantial step towards addressing one of the key current issues in the field of machine learning, namely how to select the underlying representation that an agent uses when learning.

Perhaps more importantly, Shimon's work actively brings researchers in temporal difference learning and evolutionary computation — two largely disjoint communities that focus on similar problems — closer together by being recognized and respected in both communities. This cross-disciplinary aspect of his work was the biggest risk involved, especially for a Ph.D. student who has an eye towards academia. There was a chance that the research would not be accepted by anybody. Instead, he was able to achieve recognition in both areas.

By way of the research presented in this book, Shimon has established himself as one the pre-eminent worldwide experts on machine learning for sequential decision making tasks. A particular strength of the research is its detailed empirical analysis of both the capabilities and the limitations of all variants of his proposed algorithms. In addition, Shimon's clear writing style, full explanation of background material, and detailed survey of related work make his book useful beyond its own research contributions.

In short, for both newcomers to the field and for practitioners looking for nuanced detail, this book has plenty to offer!

Austin, TX,
December 2009

Peter Stone, Associate Professor
University of Texas at Austin

Preface

This book presents the main results of the research I conducted as a Ph.D. student at The University of Texas at Austin, primarily between 2004 and 2007. The primary contributions are new algorithms for *reinforcement learning*, a form of machine learning in which an autonomous agent seeks an effective control policy for tackling a sequential decision task. Unlike in supervised learning, the agent never sees examples of correct or incorrect behavior but receives only a reward signal as feedback. One limitation of current methods is that they typically require a human to manually design a representation for the solution (e.g. the internal structure of a neural network). Since poor design choices can lead to grossly suboptimal policies, agents that automatically adapt their own representations have the potential to dramatically improve performance. This book introduces two novel approaches for automatically discovering high-performing representations.

The first approach synthesizes temporal difference methods, the traditional approach to reinforcement learning, with evolutionary methods, which can learn representations for a broad class of optimization problems. This synthesis is accomplished via 1) *on-line evolutionary computation*, which customizes evolutionary methods to the on-line nature of most reinforcement learning problems, and 2) *evolutionary function approximation*, which evolves representations for the value function approximators that are critical to the temporal difference approach.

The second approach, called *adaptive tile coding*, automatically learns representations based on tile codings, which form piecewise-constant approximations of value functions. It begins with coarse representations and gradually refines them during learning, analyzing the current policy and value function to deduce the best refinements.

This book also introduces a novel method for devising input representations. In particular, it presents a way to find a minimal set of features sufficient to describe the agent's current state, a challenge known as the *feature selection* problem. The technique, called *Feature Selective NEAT* is an extension to NEAT, a method for evolving neural networks used throughout this work. While NEAT evolves both the topology and weights of a neural network, FS-NEAT goes one step further by learning the network's inputs too. Using evolution, it automatically and simultaneously determines the network's inputs, topology, and weights.

In addition to introducing these new methods, this book presents extensive empirical results in multiple domains demonstrating that these techniques can substantially improve performance over methods with manual representations.

The research presented in this book would not have been possible without the critical contributions of many collaborators. These include Peter Stone, my Ph.D. advisor; Risto Miikkulainen, a member of my thesis committee; and my colleagues Nate Kohl, Ken Stanley, and Matt Taylor. In addition, this research was supported in part by grants from IBM, NASA, NSF, and DARPA and by an IBM Ph.D. Fellowship.

Amsterdam, The Netherlands
February 2010

Shimon Whiteson, Assistant Professor
University of Amsterdam

Contents

Chapter 1
Introduction

The goal of *reinforcement learning* (149) is to enable autonomous agents to learn effective control policies for challenging tasks. Rather than relying on directions from a human expert, a reinforcement learning agent uses its experience interacting with the world to infer a strategy for solving the given problem. Unlike supervised learning methods (96), reinforcement learning methods do not need access to examples of correct or incorrect behavior. Instead, the agent needs only a reward signal to quantify the immediate effects of its actions and it can learn a control policy to maximize the reward it accrues in the long term.

The agent's control policy is a function mapping each *state* the agent may experience to the *action* it should take in that state. Ideally, an autonomous agent would discover this policy without any human assistance, merely by learning from experience. In practice, however, current methods require substantial input from a human designer in order to perform well. The designer typically must select an appropriate learning algorithm, set parameters for that algorithm, and specify a representation for the agent's policy. This representation typically consists of the following parts:

1. the *state representation*, which could consist of low-level sensory data or high-level salient features extracted from that data,
2. the *internal representation*, which specifies a set of parameters and the way the policy is computed from those parameters, and
3. the *action representation*, which could consist of low-level actuator settings or high-level operations that require many steps to complete.

The bulk of this book focuses on automating the design of the second of these parts, the internal representation. Hence, the central question this book addresses is: given adequate representations for states and actions, how can a reinforcement learning agent automatically discover an internal representation for a control policy that maps those states to actions? This chapter discusses the motivation for addressing this question, outlines the approach taken, and briefly overviews the contents of the following chapters. Throughout the remainder of this book, "representation" refers to the agent's internal representation, unless otherwise specified.

S. Whiteson: Adaptive Representations for Reinforcement Learning, SCI 291, pp. 1–5.
springerlink.com

1.1 Motivation

Intelligent systems are adaptive by nature; hence, machine learning methods are critical to the progress of artificial intelligence. Many practical methods have been developed for supervised learning, where the agent learns from examples of correct and incorrect behavior, and have been successfully applied to a range of real-world problems, from spam filtering (7) to credit card fraud detection (34).

However, there are many important problems (e.g., robot control, game playing, and system optimization) to which supervised learning methods may not be applicable because no human expert is available to provide correctly labeled training examples or because doing so is infeasibly expensive. However, an agent can still learn to solve such problems if the human designer can describe its goal or, more generally, quantify a *reward function*. The challenge of reinforcement learning is to devise algorithms that enable an agent, while interacting with its environment, to find an effective control policy given feedback only from this reward function.

Many methods already exist for solving reinforcement learning problems. However, such methods often do not perform well in domains that are highly stochastic and/or have large or continuous state spaces. As a result, there have been relatively few successful real-world applications of reinforcement learning, e.g. (151; 38; 104).

A chief limitation of current methods is their reliance on human expertise to design critical aspects of the agent's solution. Though no labeled training examples are provided, the human designer still must determine which learning algorithm to use, how to set its parameters, and how to represent the agent's solution. For reinforcement learning methods to become more practical, they must perform well even when the expertise necessary to perform such design steps is not available. Hence, the development of new methods that automate this design process is a critical goal.

This book takes a step in that direction by introducing methods that enable a reinforcement learning agent to automatically discover effective internal representations. It also presents empirical results verifying that these methods can substantially improve performance over manually designed representations in several reinforcement learning tasks.

1.2 Approach

This book presents two fundamentally different approaches to devising adaptive representations for reinforcement learning. The first approach synthesizes temporal difference methods, the traditional approach to reinforcement learning, with evolutionary methods, which can learn representations for a broad class of optimization problems. The first step towards this synthesis is *on-line evolutionary computation*, a method which borrows exploratory mechanisms traditionally used in temporal difference methods and uses them to help evolutionary methods cope better with the *on-line* nature of most reinforcement learning problems.

Customizing evolutionary methods for on-line problems paves the way for *evolutionary function approximation*, the second step in synthesizing these two approaches. Evolutionary function approximation fully integrates temporal difference and evolutionary methods by evolving representations, not for policies, but for the value function approximators central to the temporal difference approach. Each member of the population, rather than remaining fixed during its lifetime, learns via temporal difference methods. Hence, this approach *evolves* agents that are better able to *learn*.

The resulting method is an improvement over traditional temporal difference methods because it automates the design of value function approximator representations. It is also an improvement over the traditional evolutionary approach because it 1) uses temporal difference methods to exploit the specific structure of the reinforcement learning problem and 2) enables powerful syngeries between evolution and learning, such as the *Baldwin Effect*. Furthermore, when combined with on-line evolutionary computation, this method can excel at on-line tasks.

This book also presents a variation of evolutionary function approximation designed to be more *sample-efficient*, i.e., to minimize the number of interactions with the real world required to learn a good policy. By saving experience gathered from previous generations, sample-efficient evolutionary function approximation can train each new generation off-line using only computation time: no additional sample episodes are needed. The resulting function approximators can then be evaluated and selectively reproduced in many fewer episodes.

In principle, evolutionary function approximation is applicable to any type of representation that can be evolved, though this book studies only its application to neural networks. By contrast, *adaptive tile coding*, the second approach to devising adaptive representations for reinforcement learning, is specific to one type of representation: tile coding. Tile coding is a simple, linear representation that has enjoyed considerable empirical success (144; 140). It works by dividing the state space into disjoint tiles which are used to learn a piecewise-constant value function approximation. However, it requires a human designer to correctly select the width of each tile in each dimension.

Adaptive tile coding automates this design process by starting with large tiles and making them smaller during learning by splitting existing tiles in two. Unlike neural networks, which tend to operate like "black boxes," tile codings are typically much easier to interpret: changes to the representation (e.g., splitting tiles in two) have consequences that are largely predictable. Hence, an agent, by analyzing its own behavior, can reason about how to improve its tile coding representation without the need for expensive evolution. In addition to automatically finding good representations, this approach gradually reduces the function approximator's level of generalization over time, a factor known to critically affect performance in tile coding (126).

Both evolutionary function approximation and adaptive tile coding focus on automating the design of the agent's *internal* representation. However, this book also presents a novel method for devising *state* representations. In particular, it presents a way to find a minimal set of features sufficient to describe the agent's current state, a challenge known as the *feature selection* problem (24). The technique, called *Feature Selective NEAT* (FS-NEAT) is an extension to NEAT (137), a method for evolving neural networks used throughout this book. While NEAT evolves both the topology and weights of a neural network, FS-NEAT goes one step further by learning the network's inputs too. Using evolution, it automatically and simultaneously determines the network's inputs, topology, and weights.

1.3 Overview

The remainder of this book is organized as follows. Chapter 2 provides a brief introduction to reinforcement learning. It describes the standard reinforcement learning framework and describes the two main approaches to solving reinforcement learning problems, *temporal difference* methods and *policy search* methods and details the specific base learning algorithms used throughout this book.

Chapter 3 introduces *on-line evolutionary computation*, which customizes evolutionary methods to the on-line nature of many reinforcement learning problems This chapter introduces three variations of on-line evolutionary computation and presents detailed empirical results comparing these variations to the original off-line approach in two reinforcement learning tasks: the mountain car and server job scheduling domains.

Chapter 4 describes *evolutionary function approximation*, which harnesses the representation-learning power of evolutionary methods to improve temporal difference function approximators. This chapter presents detailed empirical results in the mountain car and scheduling domains comparing this approach to 1) evolutionary methods in the absence of temporal difference methods and 2) temporal difference learning alone with a range of manually designed function approximators. It also compares the best results to other learning and non-learning approaches to these domains, compares *Darwinian* and *Lamarckian* implementations of evolutionary function approximation, and presents some additional tests that offer insight into why certain methods outperform others in these domains and what factors can make neural network function approximation difficult in practice.

Chapter 5 presents *sample-efficient* evolutionary function approximation and compares its performance to the original evolutionary function approximation method in a variation of the server job scheduling task that is designed to be deterministic, the case where sample-efficient learning is most critical.

Chapter 6 introduces *Feature Selective NEAT* (FS-NEAT) and evaluates it in RARS, a challenging automobile racing task. This chapter presents experiments comparing FS-NEAT to the original NEAT method in terms of performance as well as the size and number of inputs of the evolved networks. These experiments are repeated across a range of increasingly difficult feature selection problems by varying the number of irrelevant and redundant features available to the agent.

Chapter 7 describes two variations of *adaptive tile coding* which use different criteria for determining which tiles to split, one based on expected changes to the value function and the other based on expected changes to the policy. This chapter presents experiments comparing both versions of adaptive tile coding to various manually designed tile codings in both the mountain car and puddle world domains.

It also examines qualitative properties of the final learned policies and value functions to better understand why the methods perform as they do.

Chapter 8 surveys a broad range of previous research that is related in terms of both methods and goals to the work presented in this book. It discusses other methods for optimizing representations in supervised learning, reinforcement learning, and evolutionary computation. It also overviews other research about combining evolution and learning with applications to both supervised and reinforcement learning tasks. Furthermore, it surveys previous work on balancing exploration and exploitation, in the context of k-armed bandit problems, associative search, and reinforcement learning. Finally, this chapter discusses previous work on feature selection, surveying both *filter* and *wrapper* methods and discussing their relationship to FS-NEAT.

Chapter 9 enumerates the primary conclusions of this book, mentions some negative results obtained in the course of this research, addresses some of the broader implications of the book, and outlines ideas for future work.

Chapter 2
Reinforcement Learning

Reinforcement learning (67; 149) is a type of *machine learning* (96) in which an agent seeks an effective *policy* for solving a sequential decision task. Such a policy dictates how the agent should behave in each *state* it may encounter. Unlike *supervised learning*, the agent never sees examples of correct or incorrect behavior but instead receives only a numerical reward signal. The agent's actions affect not only the immediate reward it receives but also the next state it experiences and, consequently, future opportunities for reward. Hence, a reinforcement learning agent seeks a policy that maximizes, not the immediate reward signal, but the total reward accrued over the long term.

Reinforcement learning is an important tool in many scenarios that require adaptive agents (e.g., robot control, game playing, and system optimization). Often, the human designer does not know how an agent should behave and so cannot generate the examples necessary for supervised learning. However, if he or she can describe the agent's goal or, more generally, quantify the costs and benefits of different outcomes, then the agent can autonomously discover an effective policy via reinforcement learning. In other words, if the designer provides the reinforcement, the agent can learn to maximize it.

This chapter provides a brief introduction to reinforcement learning. First, it describes the standard reinforcement learning framework. Next, it describes two main approaches to solving reinforcement learning problems, *temporal difference* (147) and *policy search* methods, and details the specific base learning algorithms used throughout this book.

2.1 Reinforcement Learning Framework

The standard reinforcement learning framework, depicted in Figure 2.1, consists of an agent repeatedly interacting with its environment at discrete intervals (67; 149). At each timestep t, the agent perceives the environment's current state, $s_t \in S$, where S is the set of all possible states, and selects an action $a_t \in A$, where A is the set of all possible actions. The environment responds with a reward signal $r_{t+1} \in \Re$ and a

S. Whiteson: Adaptive Representations for Reinforcement Learning, SCI 291, pp. 7–15.
springerlink.com

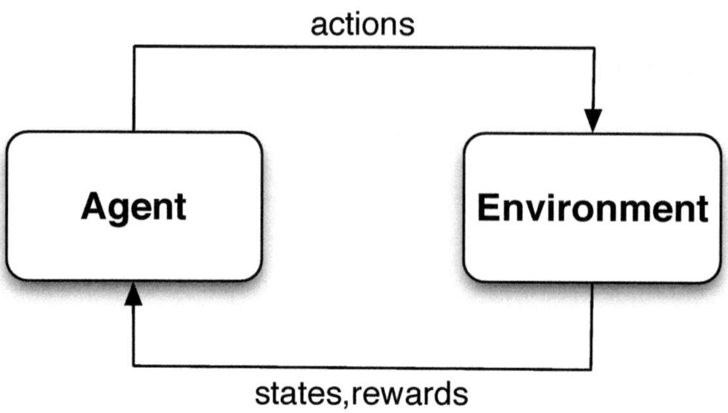

Fig. 2.1 The reinforcement learning framework, in which an agent takes a series of actions, each of which generates a reward and a new state

new state s_{t+1}. Often we assume the state space is *factored*, in which case each state is a vector of n state *features*: $s_i = \langle x_1, x_2, ..., x_n \rangle \in \Re^n$.

We also typically assume that the environment satisfies the *Markov property*, which holds when the probability that the agent perceives a given state and reward depends only on the previous state and action. In other words, the Markov property is satisfied if the following equation always holds:

$$Pr\{s_{t+1}=s', r_{t+1}=r \,|\, s_t, a_t, r_t, s_{t-1}, a_{t-1}, ..., r_1, s_0, a_0\} = Pr\{s_{t+1}=s', r_{t+1}=r \,|\, s_t, a_t\}$$

A reinforcement learning task that satisfies the Markov property is called a *Markov decision process* (MDP) (21) and can be described as a four-tuple $\langle S, A, T, R \rangle$. As before, S is the set of all states and A is the set of all actions. $T : S \times A \times S \mapsto [0,1]$ specifies the probability of transitioning to any state,

$$T(s,a,s') = Pr\{s_{t+1} = s' \,|\, s_t = s, a_t = a\}$$

and $R : S \times A \times S \mapsto \Re$ specifies the expected immediate reward,

$$R(s,a,s') = E\{r_{t+1} \,|\, s_t = s, a_t = a, s_{t+1} = s'\}$$

The goal of the agent is to maximize the long-term discounted reward it will accrue in the future, which at time t is $\sum_{k=0}^{\infty} \gamma^k r_{t+k+1}$ where $\gamma \in [0,1]$ is a discount parameter. To maximize this quantity, the agent must learn a policy $\pi : S \mapsto A$. $\pi(s)$ specifies the action the agent takes in state s. Every policy has an associated *state value function* $V^\pi : S \mapsto \Re$ which specifies the expected long-term discounted reward the agent will receive starting in state s and following policy π thereafter:

$$V^\pi(s) = E\{\sum_{k=0}^{\infty} \gamma^k r_{t+k+1} \,|\, \pi, s_t = s\}$$

Every policy also has an *action value function* $Q^\pi : S \times A \mapsto \Re$, which specifies the expected long-term discounted reward the agent will receive if it takes action a in state s and follows policy π thereafter:

$$Q^\pi(s,a) = E\{\sum_{k=0}^{\infty} \gamma^k r_{t+k+1} \mid \pi, s_t = s, a_t = a\}$$

For every MDP there exists an optimal value function V^* such that $V^*(s) = \max_\pi V^\pi(s)$, an optimal action value function, Q^* such that $Q^*(s,a) = \max_\pi Q^\pi(s,a)$, and at least one optimal policy π^* such that:

$$\pi^*(s) = \text{argmax}_a Q^*(s,a) = \text{argmax}_a \sum_{s'} T(s,a,s')[R(s,a,s') + \gamma V^*(s')]$$

The goal of a reinforcement learning agent is to find or approximate π^*. When T and R are unknown, the agent can learn only by interacting with the environment and observing state transitions and rewards. The rest of this section introduces two major approaches for doing so.

2.2 Temporal Difference Methods

Value and functions are important, not just for measuring the worth of a given policy, but for discovering good policies. In fact, many reinforcement learning algorithms do not directly search for policies at all but instead strive to find the optimal value function.

If S is finite and the agent has a model of its environment, (i.e., if T and R are known), then the optimal value function can be computed using *dynamic programming* (20). Dynamic programming works by exploiting the close relationship between consecutive states, as expressed in the Bellman optimality equation:

$$V^*(s) = \max_a \sum_{s'} T(s,a,s')[R(s,a,s') + \gamma V^*(s')]$$

This relationship means that an estimate of the value of any given state can be constructed based on estimates of the states that might occur next. This bootstrapping process is the central premise of dynamic programming and can be achieved by turning the Bellman optimality equation into an update rule. For example, *value iteration* (114) is a dynamic programming method that begins with an arbitrary value function V^0 and applies the following update rule for each $s \in S$:

$$V^{k+1}(s) = \max_a \sum_{s'} T(s,a,s')[R(s,a,s') + \gamma V^k(s')]$$

Value iteration and other dynamic programming methods are guaranteed to converge to the optimal value function when S is finite. However, their practical usefulness is limited by the assumption that a model is available.

In reinforcement learning, an agent interacts with an environment for which neither T nor R are known. As a result, dynamic programming methods are not directly applicable for two reasons. First, the value iteration update cannot be computed. Second, it is no longer sufficient to learn V^* since computing π^* from it requires knowing T and R. However, if the agent can learn Q^*, it can derive π^* from it without knowing T and R. Fortunately, Q^* can be learned without a model by using *temporal difference* methods (147), which synthesize dynamic programming with Monte Carlo methods. Each time an agent in state s_t takes an action a_t, the reward r_{t+1} it receives and the state s_{t+1} to which it transitions can be used to estimate the role of T and R in the update. For example, *Q-learning* (158), a popular temporal difference method, employs the following update rule:

$$Q(s_t,a_t) \leftarrow (1-\alpha)Q(s_t,a_t) + \alpha[r_{t+1} + \gamma \max_a Q(s_{t+1},a)]$$

where $\alpha \in [0,1]$ is a learning rate parameter. The update rule moves the old estimate $Q(s_t,a_t)$ closer to an estimated *target* $r_{t+1} + \gamma \max_a Q(s_{t+1},a)$ by an amount controlled by α.

Since T and R are unknown, temporal difference methods cannot simply iterate over S and A to perform updates. Instead, the agent can only perform updates based on transitions and rewards it observes while interacting with its environment. Like value iteration, Q-learning converges to the optimal value function when S is finite but only if the agent explores its environment in a manner that guarantees it visits every state infinitely often. Hence, temporal difference methods are typically coupled with exploration mechanisms which ensure that the agent, rather than always behaving greedily with respect to its current value function, sometimes tries alternative actions. The simplest exploration mechanism is called *ϵ-greedy exploration* (158), whereby at each timestep the agent takes a random action with probability ϵ and the greedy action otherwise.

In simple reinforcement learning tasks, the value function can be represented in a table, with one entry for each state-action pair. However, for most real-world tasks this approach is infeasible because $|S|$ grows exponentially with respect to the number of state features, a problem Bellman dubbed the "curse of dimensionality" (20). Hence, the agent may be unable to even store such a table, much less learn correct values for each entry in reasonable time. Moreover, many problems have continuous state features, in which case S is not finite and a table-based approach is impossible even in principle.

In such cases, temporal difference methods rely on *function approximation*. In this approach, the value function is not represented exactly but instead approximated via a parameterized function. Typically, those parameters are incrementally adjusted via supervised learning methods to make the function's output more closely match estimated targets generated from the agent's experience. Many different methods of function approximation have been used successfully, including tile coding, radial basis functions, and neural networks (149).

Algorithm 1 Q-LEARN($S, A, \sigma, c, \alpha, \gamma, \lambda, \varepsilon_{td}, e$)

1: *// S: set of all states, A: set of all actions, σ: standard deviation of initial weights*
2: *// c: output scale, α: learning rate, γ: discount factor, λ: eligibility decay rate*
3: *// ε_{td}: exploration rate, e: total number of episodes*
4:
5: $N \leftarrow$ INIT-NET(S, A, σ) *// make a new network N with random weights*
6: **for** $i \leftarrow 1$ to e **do**
7:     $s, s' \leftarrow$ null, INIT-STATE(S) *// environment picks episode's initial state*
8:     **repeat**
9:         $Q[] \leftarrow c \times$ EVAL-NET(N, s') *// compute value estimates for current state*
10:         **with-prob**(ε_{td}) $a' \leftarrow$ RANDOM(A) *// select random exploratory action*
11:         **else** $a' \leftarrow \text{argmax}_j Q[j]$ *// or select greedy action*
12:         **if** $s \neq$ null **then**
13:             BACKPROP($N, s, a, (r + \gamma \max_j Q[j])/c, \alpha, \gamma, \lambda$) *// adjust weights toward target*
14:         $s, a \leftarrow s', a'$
15:         $r, s' \leftarrow$ TAKE-ACTION(a') *// take action and transition to new state*
16:     **until** TERMINAL-STATE?(s)

Algorithm 1 describes the Q-learning algorithm when a neural network is used for function approximation. The inputs to the network describe the agent's current state; the outputs, one for each action, represent the agent's current estimate of the value of the associated state-action pairs. The initial weights of the network are drawn from a Gaussian distribution with mean 0.0 and standard deviation σ (line 5). The EVAL-NET function (line 9) returns the activation on the network's outputs after the given inputs are fed to the network and propagated forward. Since the network uses a sigmoid activation function, these values will all be in $[0, 1]$ and hence are rescaled according to a parameter c. At each step, the weights of the neural network are adjusted (line 13) such that its output better matches the current value estimate for the state-action pair. The adjustments are made via the BACKPROP function, which implements the standard backpropagation algorithm (121) with the addition of accumulating eligibility traces controlled by λ (147). The agent uses ε-greedy selection (lines 10–11) and interacts with the environment via the TAKE-ACTION function (line 15), which returns a reward and a new state.

By addressing large and continuous state spaces, function approximation can greatly extend the applicability of temporal difference methods. However, using function approximators successfully in practice requires making crucial representational decisions, e.g., choosing the number of hidden units and initial weights of a neural network. Much of this book focuses on simplifying these decisions via methods that automatically discover effective function approximator representations (see Chapters 4, 5, and 7).

2.3 Policy Search Methods

Dynamic programming and temporal difference methods rely heavily on the notion of value functions for solving reinforcement learning problems. By contrast, policy search methods do not use value functions at all. Instead, they use optimization techniques (e.g., gradient methods (145; 105; 71) or evolutionary methods (101; 172; 137)) to directly search the space of policies for one that accrues maximal reward. To assess the performance of each candidate policy, the agent typically employs the policy for one or more episodes and sums the total reward received.

Among the most successful approaches to policy search is *neuroevolution* (172), which uses *evolutionary computation* (52) to optimize a population of neural networks. In a typical neuroevolutionary system, the weights of a neural network are strung together to form an individual genome. A population of such genomes is then evolved by evaluating each one and selectively reproducing the fittest individuals through crossover and mutation. Most neuroevolutionary systems require the designer to manually determine the network's representation (i.e., how many hidden nodes there are and how they are connected).

However, some neuroevolutionary methods can automatically evolve representations along with network weights. In particular, NeuroEvolution of Augmenting Topologies (NEAT) (137) combines the usual search for network weights with evolution of the network structure.[1] It has amassed numerous empirical successes on difficult reinforcement learning tasks like non-Markovian double pole balancing (137), game playing (139), robot control (138; 150), and data filtering in high energy physics (1; 166). In reinforcement learning tasks, the networks that NEAT evolves have a similar configuration to those used by Q-learning in Algorithm 1: there is one input for each state feature, one output for each action, and the agent takes the action whose corresponding output has the highest activation. However, since the network represents a policy, not a value function, the activations on the output nodes do not represent value estimates. In fact, the outputs can have arbitrary activations so long as the most desirable action has the largest activation.

Algorithm 2 contains a high-level description of the NEAT algorithm applied to an episodic reinforcement learning problem. NEAT begins by creating a population of random networks (line 4). In each generation, NEAT repeatedly iterates over the current population (lines 6–7). During each step of a given episode, the agent takes whatever action corresponds to the output with the highest activation (lines 10–12). NEAT maintains a running total of the reward accrued by the network during its evaluation (line 13). Each generation ends after e episodes, at which point each network's average fitness is $N.fitness/N.episodes$. In stochastic domains, e typically must be much larger than $|P|$ to ensure accurate fitness estimates for each network. NEAT creates a new population by repeatedly calling the BREED-NET function (line 18), which performs crossover on two highly fit parents. The new resulting network can then undergo mutations that add nodes or links to its structure (lines 19–20). The remainder of this section provides an overview of the reproductive process that occurs in lines 17–20. Stanley and Miikkulainen (137) present a full description.

Unlike other systems that evolve network topologies and weights (57; 172) NEAT begins with a uniform population of simple networks with no hidden nodes and

[1] Parts of the following description were adapted from the original NEAT paper (137).

Algorithm 2 NEAT(S, A, p, m_n, m_l, g, e)

// S: set of all states, A: set of all actions, p: population size, m_n*: node mutation rate*
// m_l*: link mutation rate, g: number of generations, e: episodes per generation*

$P[] \leftarrow$ INIT-POPULATION(S,A, p) *// create new population P with random networks*
for $i \leftarrow 1$ to g **do**
  for $j \leftarrow 1$ to e **do**
    $N, s, s' \leftarrow P[j \% p]$, null, INIT-STATE($S$) *// select next network*
    repeat
      $Q[] \leftarrow$ EVAL-NET(N, s') *// evaluate selected network on current state*
      $a' \leftarrow \text{argmax}_i Q[i]$ *// select action with highest activation*
      $s, a \leftarrow s', a'$
      $r, s' \leftarrow$ TAKE-ACTION(a') *// take action and transition to new state*
      $N.fitness \leftarrow N.fitness + r$ *// update total reward accrued by N*
    until TERMINAL-STATE?(s)
    $N.episodes \leftarrow N.episodes + 1$ *// update total number of episodes for N*
  $P'[] \leftarrow$ new array of size p *// new array will store next generation*
  for $j \leftarrow 1$ to p **do**
    $P'[j] \leftarrow$ BREED-NET($P[]$) *// make a new network based on fit parents in P*
    with-probability m_n: ADD-NODE-MUTATION($P'[j]$) *// add node to new network*
    with-probability m_l: ADD-LINK-MUTATION($P'[j]$) *// add link to new network*
  $P[] \leftarrow P'[]$

inputs connected directly to outputs. New structure is introduced incrementally via two special mutation operators. Figure 2.2 depicts these operators, which add new hidden nodes and links to the network. Only the structural mutations that yield performance advantages tend to survive evolution's selective pressure. In this way, NEAT tends to search through a minimal number of weight dimensions and find an appropriate complexity level for the problem.

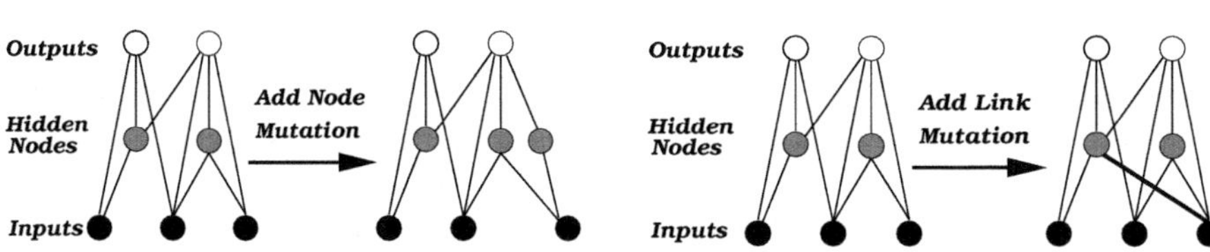

(a) A mutation operator for adding new nodes

(b) A mutation operator for adding new links

Fig. 2.2 Examples of NEAT's mutation operators for adding structure to networks. In (a), a hidden node is added by splitting a link in two. In (b), a link, shown with a thicker black line, is added to connect two nodes.

Evolving network structure requires a flexible genetic encoding. Each genome in NEAT includes a list of *connection genes*, each of which refers to two *node genes* being connected. Each connection gene specifies the in-node, the out-node, the weight of the connection, whether or not the connection gene is expressed (an enable bit), and an *innovation number*, which allows NEAT to find corresponding genes during crossover.

In order to perform crossover, the system must be able to tell which genes match up between *any* individuals in the population. For this purpose, NEAT keeps track of the historical origin of every gene. Whenever a new gene appears (through structural mutation), a *global innovation number* is incremented and assigned to that gene. The innovation numbers thus represent a chronology of every gene in the system. Whenever these genomes cross over, innovation numbers on inherited genes are preserved. Thus, the historical origin of every gene in the system is known throughout evolution.

Through innovation numbers, the system knows exactly which genes match up with which. Genes that do not match are either *disjoint* or *excess*, depending on whether they occur within or outside the range of the other parent's innovation numbers. When crossing over, the genes in both genomes with the same innovation numbers are lined up. Genes that do not match are inherited from the more fit parent, or if they are equally fit, from both parents randomly. Historical markings allow NEAT to perform crossover without expensive topological analysis. Genomes of different organizations and sizes stay compatible throughout evolution, and the problem of matching different topologies (118) is essentially avoided.

In most cases, adding new structure to a network initially reduces its fitness. However, NEAT speciates the population, so that individuals compete primarily within their own species rather than with the population at large. Hence, topological innovations are protected and have time to optimize their structure before competing with other niches in the population.

Historical markings make it possible for the system to divide the population into species based on topological similarity. The distance δ between two network encodings is a simple linear combination of the number of excess (E) and disjoint (D) genes, as well as the average weight differences of matching genes ($\overline{W}$):

$$\delta = \frac{c_1 E}{N} + \frac{c_2 D}{N} + c_3 \cdot \overline{W}$$

The coefficients c_1, c_2, and c_3 adjust the importance of the three factors, and the factor N, the number of genes in the larger genome, normalizes for genome size. Genomes are tested one at a time; if a genome's distance to a randomly chosen member of the species is less than δ_t, a compatibility threshold, it is placed into this species. Each genome is placed into the first species where this condition is satisfied, so that no genome is in more than one species. The reproduction mechanism for NEAT is *explicit fitness sharing* (52), where organisms in the same species must share the fitness of their niche, preventing any one species from taking over the population.

Evolutionary methods such as NEAT assess the value of entire policies, rather than reasoning about the value of particular state-action pairs. The holistic nature of this approach is sometimes criticized. For example, Sutton and Barto write:

> Evolutionary methods do not use the fact that the policy they are searching for is a function from states to actions; they do not notice which states an individual passes through during its lifetime, or which actions it selects. In some cases this information can be misleading (e.g., when states are misperceived) but more often it should enable more efficient search (149, p. 9).

In some contexts, these facts put evolutionary methods at a theoretical disadvantage. For example, in some circumstances dynamic programming methods are guaranteed to find an optimal policy in time polynomial in the number of states and actions (85). By contrast, evolutionary methods, in the worst case, must iterate over an exponential number of candidate policies before finding the best one.

However, in practice, evolutionary methods have proven quite effective and at least sometimes outperform temporal difference methods (137; 165). There are many possible explanations for these results, such as the ability of such methods to cope with non-Markovian environments or the fact that policies are sometimes simpler to represent than value functions. But perhaps most critical is the ability of methods like NEAT to automatically discover effective representations. Much of this book focuses on new ways of harnessing this ability, by altering evolutionary methods to make them more suitable for reinforcement learning (Chapter 3), synthesizing them with temporal difference methods so as to evolve representations for value functions (Chapters 4 and 5), or extending them to automatically select useful state features (Chapter 6).

Chapter 3
On-Line Evolutionary Computation

Sutton and Barto's criticism of evolutionary methods rests on the fact that such methods do not exploit the specific structure of the reinforcement learning problem. Instead, they just treat it like any other optimization problem, using total reward accrued as a fitness function. Much of this book focuses on eliminating this shortcoming by customizing such techniques to the unique characteristics of the reinforcement learning problem. As a result, the representation-learning power of methods like NEAT can be harnessed without sacrificing the advantages of other reinforcement learning approaches, such as temporal difference methods. The heart of this customization is presented in Chapter 4, which describes how to synthesize evolutionary and temporal difference methods so as to evolve representations for value functions.

Before doing so, however, this chapter describes how to customize evolutionary methods to the *on-line* nature of many reinforcement learning problems. While methods like NEAT have excelled on many challenging reinforcement learning problems, their empirical success is largely restricted to *off-line* scenarios, in which the agent learns, not in the real-world, but in a "safe" environment like a simulator. This chapter introduces methods that make it possible to harness the representation-learning capacity of methods like NEAT in on-line scenarios, where an agent interacts with the real world and adjusts its policy as it goes.

In off-line scenarios, an agent's only goal is to learn a good policy as quickly as possible. It does not care how much reward it accrues *while it is learning* because those rewards are only hypothetical and do not correspond to real-world costs. If the agent tries disastrous policies, only computation time is lost. At any point during learning, the performance of an off-line agent is simply the quality of the best policy it has found so far.

Unfortunately, many reinforcement learning problems cannot be solved off-line because no simulator is available. Sometimes the dynamics of the task are unknown, e.g., when a robot explores an unfamiliar environment or a chess player plays a new opponent. Other times, the dynamics of the task are too complex to accurately simulate, e.g., user behavior on a large computer network or the noise in a robot's sensors and actuators.

S. Whiteson: Adaptive Representations for Reinforcement Learning, SCI 291, pp. 17–30.
springerlink.com

In such domains, the agent has no choice but to learn on-line. In an on-line learning scenario, it is not enough for an agent to learn a good policy quickly. It must also maximize the reward it accrues while it is learning because those rewards correspond to real-world costs. For example, if a robot learning on-line tries a policy that causes it to drive off a cliff, then the negative reward the agent receives is not hypothetical; it corresponds to the very real cost of fixing or replacing the robot.[1]

To measure the performance of an on-line agent it is essential to consider the quality of the policy *currently in use*, which may be different from the best policy discovered so far. Since the agent is interacting with the real-world, it must be "charged" for each policy or action it tries. In the context of evolutionary methods, this means examining the average performance of the entire population, not just the generation champion. The goal of the agent is to maximize the total reward accrued during learning, i.e., the area under a typical learning curve.

To excel in on-line scenarios, a learning algorithm must effectively balance two competing objectives. The first objective is *exploration*, in which the agent tries alternatives to its current best policy in the hopes of improving it. The second objective is *exploitation*, in which the agent follows the current best policy in order to maximize the reward it receives.

Exploitation is important because, in practice, on-line learning problems have a *finite horizon*, which means reward can be accrued for only a limited time and learning must occur during that same time. For example, an autonomous robot gathering rocks on Mars can accrue reward only until its parts wear out. If the agent simply explores, it may discover a great policy, i.e., how to find the best rocks. However, unless a similar robot will be deployed in the same region in the future, this policy is not useful after the robot stops working. Hence, the agent must exploit in order to maximize the reward accrued before time expires.

Evolutionary methods already strive to balance exploration and exploitation. In fact, Holland (62) argues that the reproduction mechanism encourages exploration, since crossover and mutation result in novel genomes, but also encourages exploitation, since each new generation is based on the fittest members of the last one. However, reproduction allows evolutionary methods to balance exploration and exploitation only *across* generations, not *within* them. Once the members of each generation have been determined, they all typically receive the same evaluation time.

This approach makes sense in deterministic domains, where each member of the population can be accurately evaluated in a single episode. However, most real-world domains are stochastic, in which case fitness evaluations must be averaged over many episodes. In these domains, giving the same evaluation time to each

[1] The term *on-line learning* is sometimes used in a very different way: to refer to *non-stationary* learning problems where the agent's environment is changing in ways that alter the optimal policy. In such problems, the agent must continually adapt to perform well. The problems of non-stationary learning and on-line learning (as the term is used here) are orthogonal. A learning scenario can be stationary but on-line, as when an agent trains in a static but real-world environment. A learning scenario can also be non-stationary but off-line, as when an agent trains on a simulator that is being continually refined. This book does not address non-stationary learning problems.

member of the population can be grossly suboptimal because, within a generation, it is purely exploratory. Instead, an on-line evolutionary algorithm should exploit the information gained earlier in the generation to systematically give more evaluations to the most promising individuals and avoid re-evaluating the weakest ones. Doing so allows evolutionary methods to increase the reward accrued during learning.

This chapter presents a novel approach, called *on-line evolutionary computation* (161; 162), which strives to achieve this balance. Instead of giving each individual the same number of episodes, on-line evolutionary computation exploits the information gained from early episodes to favor the most promising candidate policies and thereby boost the reward accrued during learning. This method works by borrowing action selection mechanisms traditionally used in temporal difference methods and applying them in evolutionary computation. TD methods naturally excel in on-line scenarios because they use action selection mechanisms to control how often the agent exploits (by behaving greedily with respect to current value estimates) and how often it explores (by trying alternative actions). This chapter describes ways to borrow the selection mechanisms used by TD methods to choose individual actions and use them in evolution to choose policies for evaluation. This approach enables evolution to excel on-line by balancing exploration and exploitation within *and* across generations.

In a sense, the problem faced by evolutionary methods is the opposite of that faced by TD methods. Within each generation, evolutionary methods naturally explore, by evaluating each member of the population equally, and so need a way to force more exploitation. By contrast, TD methods naturally exploit, by following the greedy policy, and so need a way to force more exploration. However, the goal is the same: a proper balance between the two extremes.

To apply TD action selection mechanisms in evolutionary computation, we must modify the level at which selection is performed. Evolutionary algorithms cannot perform selection at the level of individual actions because, lacking value functions, they have no notion of the value of individual actions. However, they can perform selection at the level of episodes, in which entire policies are assessed holistically. The same selection mechanisms used to choose individual actions in TD methods can be used to select policies for evaluation, allowing evolution to excel on-line by balancing exploration and exploitation within and across generations.

This chapter investigates three methods based on this approach. The first, based on ϵ-greedy selection (158), switches probabilistically between searching for better policies and re-evaluating the best known policy. The second, based on softmax selection (149), distributes evaluations in proportion to each individual's estimated fitness. The third, based on interval estimation (66), computes confidence intervals for the fitness of each policy and always evaluates the policy with the highest upper bound.

These methods were evaluated by implementing them in NEAT and testing their performance in two domains: 1) mountain car, a canonical reinforcement learning benchmark task, and 2) server job scheduling, a large stochastic reinforcement learning task from the field of *autonomic computing* (69). The results demonstrate that these techniques can substantially improve the on-line performance of evolutionary methods and that softmax selection and interval estimation are more effective than the simple ε-greedy approach. As a result, the ability of NEAT to discover effective representations can be harnessed, not just in off-line scenarios, but in on-line scenarios too.

3.1 ε-Greedy Evolution

When ε-greedy selection is used in TD methods, a single parameter ε controls what fraction of the time the agent deviates from greedy behavior. Each time the agent selects an action, it chooses probabilistically between exploration and exploitation. With probability ε, it explores by selecting randomly from the available actions. With probability $1-\varepsilon$, it exploits by selecting the greedy action.

In evolutionary computation, this same mechanism can be used at the beginning of each episode to select a policy for evaluation. With probability ε, the algorithm selects a policy randomly. With probability $1-\varepsilon$, the algorithm exploits by selecting the best policy discovered so far in the current generation. The score of each policy is just the average reward per episode it has received so far. Each time a policy is selected for evaluation, the total reward it receives is incorporated into that average, which can cause it to gain or lose the rank of best policy.

To apply ε-greedy selection to NEAT, we need only alter the way networks are selected for evaluation. Instead of iterating through the population repeatedly until e episodes are complete (lines 6–7 in Algorithm 2), NEAT selects for evaluation, at the beginning of each episode, the policy returned by the ε-greedy selection function described in Algorithm 3. This function returns a policy p which is either selected randomly or which maximizes $f(p)$, the fitness of p averaged over all the episodes for which it has been previously evaluated.

Algorithm 3 ε-GREEDY SELECTION(P,ε)

// P: population, ε: NEAT's exploration rate

with-prob(ε) return RANDOM(P) *// select random member of population*
else return $\text{argmax}_{p \in P} f(p)$ *// or select current generation champion*

Using ε-greedy selection in evolutionary computation allows it to thrive in on-line scenarios by balancing exploration and exploitation. For the most part, this method does not alter evolution's search but simply interleaves it with exploitative episodes that increase average reward during learning. The next section describes how softmax selection can be applied to evolution to create a more nuanced balance between exploration and exploitation.

3.2 Softmax Evolution

When softmax selection is used in TD methods, an action's probability of selection is a function of its estimated value. In addition to ensuring that the greedy action is chosen most often, this technique focuses exploration on the most promising alternatives. There are many ways to implement softmax selection but one popular method relies on a Boltzmann distribution (149), in which case an agent in state s chooses an action a with probability

$$\frac{e^{Q(s,a)/\tau}}{\sum_{a' \in A} e^{Q(s,a')/\tau}} \tag{3.1}$$

where $\tau \in [0, \infty]$ is a parameter controlling the degree to which actions with higher values are favored in selection. The higher the value of τ, the more equiprobable the actions are.

As with ε-greedy selection, we can use softmax selection in evolution to select policies for evaluation. At the beginning of each generation, each individual is evaluated for one episode, to initialize its fitness. Then, the remaining $e - |P|$ episodes are allocated according to a Boltzmann distribution. Before each episode, a policy $p \in P$ is selected with probability

$$\frac{e^{f(p)/\tau}}{\sum_{p' \in P} e^{f(p')/\tau}} \tag{3.2}$$

where $f(p)$ is the fitness of policy p, averaged over all the episodes for which it has been previously evaluated. In NEAT, softmax selection is applied in the same way as ε-greedy selection, except that the policy selected for evaluation is that returned by the softmax selection function described in Algorithm 4, where $e(p)$ is the total number of episodes for which a policy p has been evaluated so far.

Algorithm 4 SOFTMAX SELECTION(P, τ)

// P: population, τ: softmax temperature

if $\exists\, p \in P \mid e(p) = 0$ **then**
  return p *// give each policy one evaluation first*
else
  $total \leftarrow \sum_{p \in P} e^{f(p)/\tau}$ *// compute denominator in Boltzmann expression*
  for all $p \in P$ **do**
    with-prob($\frac{e^{f(p)/\tau}}{total}$) return p *// decide whether to select p*
    else $total \leftarrow total - e^{f(p)/\tau}$ *// if not, adjust denominator*

Softmax selection provides a more nuanced balance between exploration and exploitation than ε-greedy because it focuses its exploration on the most promising alternatives to the current best policy. Softmax selection can quickly abandon poorly performing policies and prevent them from reducing the reward accrued during learning.

3.3 Interval Estimation Evolution

An important disadvantage of both ϵ-greedy and softmax selection is that they do not consider the uncertainty of the estimates on which they base their selections. One approach that addresses this shortcoming is interval estimation (66). When used in TD methods, interval estimation computes a $(100-\alpha)\%$ confidence interval for the value of each available action. The agent always takes the action with the highest upper bound on this interval. This strategy favors actions with high estimated value and also focuses exploration on the most promising but uncertain actions. The α parameter controls the balance between exploration and exploitation, with smaller values generating greater exploration.

The same strategy can be employed within evolution to select policies for evaluation. At the beginning of each generation, each individual is evaluated for one episode, to initialize its fitness. Then, the remaining $e-|P|$ episodes are allocated to the policy that currently has the highest upper bound on its confidence interval. In NEAT, interval estimation is applied just as in ϵ-greedy and softmax selection, except that the policy selected for evaluation is that returned by the interval estimation function described in Algorithm 5, where $[0,z(x)]$ is an interval within which the area under the standard normal curve is x. $f(p)$, $\sigma(p)$ and $e(p)$ are the fitness, standard deviation, and number of episodes, respectively, for policy p.

Algorithm 5 INTERVAL ESTIMATION(P,α)

```
// P: population, α: uncertainty in confidence interval

if ∃ p ∈ P | e(p) = 0 then
    return p
else
    return argmax_{p∈P}[f(p) + z((100−α)/200) σ(p)/√e(p)]
```

$$\text{return } \operatorname{argmax}_{p\in P}\left[f(p)+z\left(\frac{100-\alpha}{200}\right)\frac{\sigma(p)}{\sqrt{e(p)}}\right]$$

3.4 Testbed Domains

The methods described above were tested in two different reinforcement learning domains. The first domain, mountain car, is a standard reinforcement learning benchmark task. The second domain, server job scheduling, is a large, stochastic domain from the field of autonomic computing.

3.4.1 Mountain Car

In the mountain car task (28), depicted in Figure 3.1, an agent strives to drive a car to the top of a steep mountain. The car cannot simply accelerate forward because its engine is not powerful enough to overcome gravity. Instead, the agent must learn to drive backwards up the hill behind it, thus building up sufficient inertia to ascend to the goal before running out of speed.

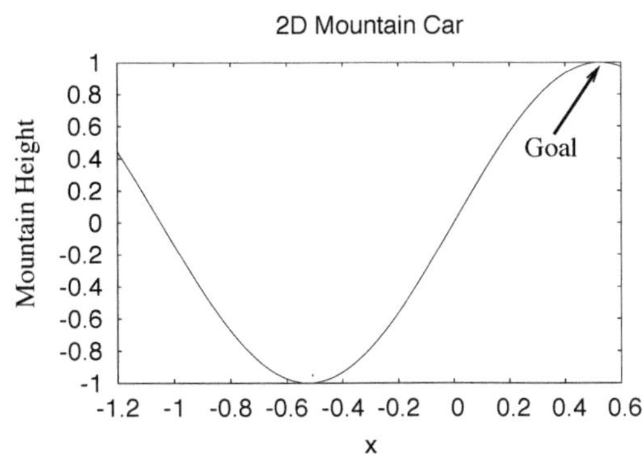

Fig. 3.1 The mountain car task, in which an underpowered car strives to reach the top of a mountain

The agent's state at timestep t consists of its current position p_t and its current velocity v_t. It receives a reward of -1 at each time step until reaching the goal, at which point the episode terminates. The agent's three available actions correspond to the throttle settings $1, 0$, and -1. The following equations control the car's movement:

$$p_{t+1} = bound_p(p_t + v_{t+1})$$

$$v_{t+1} = bound_v(v_t + 0.001a_t - 0.0025cos(3p_t))$$

where a_t is the action the agent takes at timestep t, $bound_p$ enforces $-1.2 \leq p_{t+1} \leq 0.5$, and $bound_v$ enforces $-0.07 \leq v_{t+1} \leq 0.07$. In each episode, the agent begins in a state chosen randomly from these ranges. To prevent episodes from running indefinitely, each episode is terminated after 2,500 steps if the agent still has not reached the goal.

To represent the agent's current state to the network, each state feature is divided into ten regions. One input was associated with each region (for a total of twenty inputs) and was set to one if the agent's current state fell in that region and to zero otherwise. Hence, only two inputs were activated for any given state. The agent's state could be represented more compactly, using one real-valued input for position and another for velocity. However, informal experiments found that this representation did not perform as well. The networks have three outputs, each corresponding to one of the actions available to the agent.

3.4.2 Server Job Scheduling

While the mountain car task is a useful benchmark, it is a very simple domain. To assess whether on-line evolutionary computation can scale to a much more complex problem, a challenging reinforcement learning task called server job scheduling was used. This domain is drawn from the burgeoning field of autonomic computing (69). The goal of autonomic computing is to develop computer systems that automatically configure themselves, optimize their own behavior, and diagnose and repair their own failures. The demand for such features is growing rapidly, since computer systems are becoming so complex that maintaining them with human support staff is increasingly infeasible.

The vision of autonomic computing poses new challenges to many areas of computer science, including architecture, operating systems, security, and human-computer interfaces. However, the burden on artificial intelligence is especially great, since intelligence is a prerequisite for self-managing systems. In particular, machine learning will likely play a primary role, since computer systems must be adaptive if they are to perform well autonomously. There are many ways to apply supervised methods to autonomic systems, e.g., for intrusion detection (46), spam filtering (39), or system configuration (169). However, there are also many tasks where no human expert is available and reinforcement learning is applicable, e.g network routing (27), job scheduling (160), and cache allocation (53).

One such task is server job scheduling, in which a server, such as a website's application server or database, must determine in what order to process the jobs currently waiting in its queue. Its goal is to maximize the aggregate utility of all the jobs it processes. A *utility function* (not to be confused with a TD value function) for each job type maps the job's completion time to the utility derived by the user (157). The problem of server job scheduling becomes challenging when these utility functions are nonlinear and/or the server must process multiple types of jobs. Since selecting a particular job for processing necessarily delays the completion of all other jobs in the queue, the scheduler must weigh difficult trade-offs to maximize aggregate utility. Also, this domain is challenging because it is large (the size of both the state and action spaces grow in direct proportion to the size of the queue) and probabilistic (the server does not know what type of job will arrive next).

The server job scheduling task is quite different from traditional scheduling tasks (173; 174). In the latter case, there are typically multiple resources available and each job has a partially ordered list of resource requirements. Server job scheduling is simpler because there is only one resource (the server) and all jobs are independent of each other. However, it is more complex in that performance is measured via arbitrary utility functions, whereas traditional scheduling tasks aim solely to minimize completion times.

Our experiments were conducted in a Java-based simulator. The simulation begins with 100 jobs preloaded into the server's queue and ends when the queue becomes empty. During each timestep, the server removes one job from its queue and completes it. During each of the first 100 timesteps, a new job of a randomly selected type is added to the end of the queue. Hence, the agent must make decisions about which job to process next even as new jobs are arriving. Since one job is processed at each timestep, each episode lasts 200 timesteps. For each job that completes, the scheduling agent receives an immediate reward determined by that job's utility function.

Four different job types were used in our experiments. Hence, the task can generate 4^{200} unique episodes. Utility functions for the four job types are shown in Figure 3.2. Users who create jobs of type #1 or #2 do not care about their jobs' completion times so long as they are less than 100 timesteps. Beyond that, they get increasingly unhappy. The rate of this change differs between the two types and switches at timestep 150. Users who create jobs of type #3 or #4 want their jobs completed as quickly as possible. However, once the job becomes 100 timesteps old, it is too late to be useful and they become indifferent to it. As with the first two job types, the slopes for job types #3 and #4 differ from each other and switch, this time at timestep 50. Note that all these utilities are negative functions of completion time. Hence, the scheduling agent strives to bring aggregate utility as close to zero as possible.

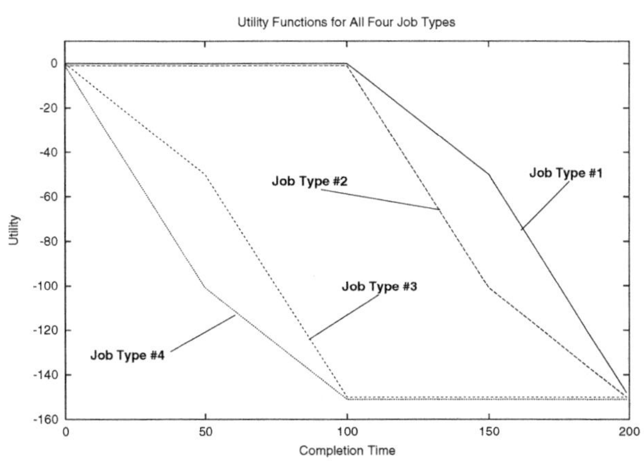

Fig. 3.2 The four utility functions used in the server job scheduling task

A primary obstacle to applying reinforcement learning methods to this domain is the size of the state and action spaces. A complete state description includes the type and age of each job in the queue. The scheduler's actions consist of selecting jobs for processing; hence a complete action space includes every job in the queue. These spaces were discretized to make them more manageable. The range of job ages from 0 to 200 is divided into four sections and the scheduler is told, at each timestep, how many jobs in the queue of each type fall in each range, resulting in 16 state features.

The action space is similarly discretized. Instead of selecting a particular job for processing, the scheduler specifies what type of job it wants to process and which of the four age ranges that job should lie in, resulting in 16 distinct actions. The server processes the youngest job in the queue that matches the type and age range specified by the action.

These discretizations mean the agent has less information about the contents of the job queue. However, its state is still sufficiently detailed to allow effective learning. Although the utility functions can change dramatically within each age range, their slopes do not change. It is the slope of the utility function, not the utility function itself, which determines how much utility is lost by delaying a given job.

The server job scheduling domain is a perfect example of a reinforcement learning task that needs to be solved on-line. Though a simulator is used for the purpose of experimental research, creating an accurate simulator in the real world would not be practical. Such a simulator would have to precisely model the server's internal workings and the behavior of all the system's users, including how that behavior changes in response to different scheduling policies. Hence, good policies can probably only be learned on-line, by trying them out on real servers. In such scenarios, maximizing on-line performance is critical, since lost reward corresponds to delays for real users.

3.5 Results

As a baseline of comparison, we applied the original, off-line version of NEAT to both the mountain car and server job scheduling domains and averaged its performance over 25 runs. The population size $|P|$ was 100 and the number of episodes per generation e was 10,000. Hence, each member of the population was evaluated for 100 episodes. Table 3.1 provides more details on the NEAT parameters used in our experiments. Next, we applied the ε-greedy, softmax, and interval estimation versions of NEAT to both domains using the same parameter settings. Each of these on-line methods has associated with it one additional parameter which controls the balance between exploration and exploitation. For each method, we experimented informally with approximately ten different settings of these parameters to find ones that worked well in the two tasks. Finally, we averaged the performance of each method over 25 runs using the best known parameter settings.

Those settings were as follows. For ε-greedy, ε was set to 0.25. This value is larger than is typically used in TD methods but makes intuitive sense, since exploration in NEAT is safer than in TD methods. After all, even when NEAT explores, the policies it selects are not drawn randomly from policy space. On the contrary, they are the children of the previous generation's fittest parents. For softmax, the appropriate value of τ depends on the range of fitness scores, which differs dramatically between the two domains. Hence, different values were required for the two domains: we set τ to 50 in mountain car and 500 in server job scheduling. For interval estimation, α was set to 20, resulting in 80% confidence intervals.

Table 3.1 The NEAT parameters used in the experiments described in this chapter. Stanley and Miikkulainen (137) describe the semantics of these parameters in detail.

Parameter	Value	Parameter	Value	Parameter	Value
weight-mut-power	0.5	recur-prop	0.0	disjoint-coeff (c_1)	1.0
excess-coeff (c_2)	1.0	mutdiff-coeff (c_3)	2.0	compat-threshold	3.0
age-significance	1.0	survival-thresh	0.2	mutate-only-prob	0.25
mutate-link-weights-prob	0.9	mutate-add-node-prob (m_n)	0.02	mutate-add-link-prob (m_l)	0.1
interspecies-mate-rate	0.01	mate-multipoint-prob	0.6	mate-multipoint-avg-prob	0.4
mate-singlepoint-prob	0.0	mate-only-prob	0.2	recur-only-prob	0.0
pop-size (p)	100	dropoff-age	100	newlink-tries	50
babies-stolen	0	num-compat-mod	0.3	num-species-target	6

Figure 3.3 summarizes the results of these experiments by plotting a uniform moving average over the last 1,000 episodes of the total reward accrued per episode for each method. We plot average reward because it is an on-line metric: it measures the amount of reward the agent accrues while it is learning. The best policies discovered by evolution, i.e. the generation champions, perform substantially higher than this average. However, using their performance as an evaluation metric would ignore the on-line cost that was incurred by evaluating the rest of population and receiving less reward per episode. Figure 3.5 plots, for the same experiments, the total cumulative reward accrued by each method over the entire run. In both graphs, error bars indicate 95% confidence intervals and Student's t-tests confirm, with 95% confidence, the statistical significance of the performance difference between each pair of methods except between softmax and interval estimation.

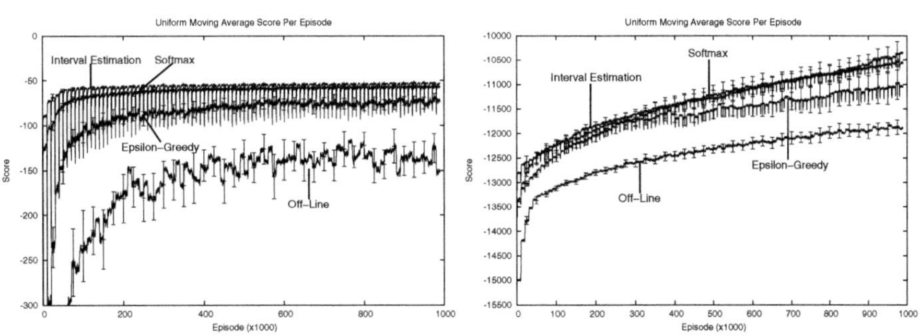

(a) Mountain Car (b) Server Job Scheduling

Fig. 3.3 The uniform moving average reward accrued by off-line NEAT, compared to three versions of on-line NEAT in the mountain car and server job scheduling domains. In both domains, all rewards are negative so the agents strive to get average reward as close to zero as possible.

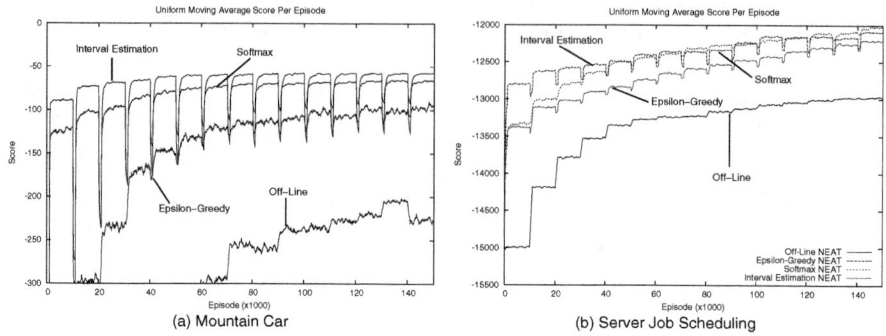

Fig. 3.4 A close-up of the early part of learning, showing the uniform moving average reward accrued by each method. Intervals corresponding to each generation of evolution are evident at this scale.

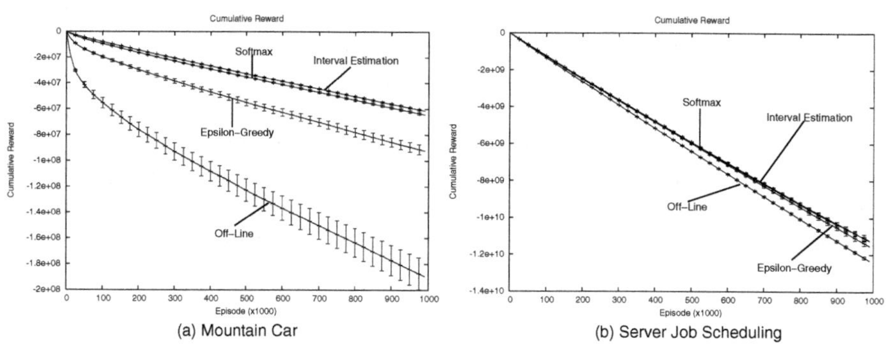

Fig. 3.5 The cumulative reward accrued by off-line NEAT, compared to three versions of on-line NEAT in the mountain car and server job scheduling domains. In both domains, all rewards are negative so the agents strive to keep cumulative reward as close to zero as possible.

3.6 Discussion

The results shown in Figure 3.3 clearly demonstrate that selection mechanisms borrowed from TD methods can dramatically improve the on-line performance of evolutionary computation. All three on-line methods substantially outperform the off-line version of NEAT. In addition, the more nuanced strategies of softmax and interval estimation fare better than ϵ-greedy. This result is not surprising since the ϵ-greedy approach simply interleaves the search for better policies with exploitative episodes that employ the best known policy. Softmax selection and interval estimation, by contrast, concentrate exploration on the most promising alternatives. Hence, they spend fewer episodes on the weakest individuals and achieve better performance as a result.

For the on-line methods, particularly interval estimation, evolution consists of a series of 10,000-episode intervals. These intervals are especially evident in Figure 3.4, which shows a close-up of the early part of learning. Each of these intervals corresponds to one generation. The performance improvements within each generation reflect the on-line methods' ability to exploit the information gleaned from earlier episodes. As the generation progresses, these methods become better informed about which individuals to favor when exploiting and average reward increases as a result.

While these intervals reveal an important feature of the on-line methods' behavior, they can make it difficult to compare performance. For example, in the mountain car domain, interval estimation begins each generation with a lot of exploration and, consequently, relatively poor performance. However, that exploration quickly pays off and its average performance rises slightly above that of softmax. Which of these two methods is receiving more reward overall? It is difficult to tell from plots of average reward. However, plots of cumulative reward, shown in Figure 3.5, are more revealing in this respect. Not surprisingly, the off-line version of NEAT accumulates much less reward than the on-line methods and ε-greedy accumulates less reward than the other on-line approaches. These graphs also show that, in mountain car, interval estimation's exploration early in each generation pays off, as it earns at least as much reward overall as softmax.

Together, these results demonstrate that borrowing selection mechanisms from TD methods can greatly improve the on-line performance of evolutionary computation. However, they do not address how on-line evolution affects the quality of the best policies discovered. Does excelling at on-line metrics necessarily hurt performance on off-line metrics? To answer this question, we selected the best policies discovered by each method (i.e. the final generation champions) and evaluated them each for 1,000 additional episodes.

In mountain car, using on-line evolution has no noticeable effect: the best policies of off-line and all three versions of on-line NEAT receive an average score of approximately -52, which matches the best results achieved in previous research on this domain (129; 144). While the mountain car domain is simple enough that all the methods find approximately optimal policies, the same is not true in scheduling, where ε-greedy performs substantially worse. Its best policies receive an average score of approximately $-11{,}100$, whereas off-line and the other two versions of on-line NEAT all receive an average score of approximately $-10{,}100$. This result is not surprising: since ε-greedy evolution spends most of its episodes re-evaluating the best policy, its fitness estimates for the rest of the population are less accurate. By focusing exploration on the most promising individuals, softmax and interval estimation offer the best of both worlds: they excel at the on-line metrics without sacrificing the quality of the best policies discovered.

Overall, these results verify the efficacy of these methods of on-line evolution. It is less clear, however, which strategy is most useful. Softmax clearly outperforms ε-greedy but may be more difficult to use in practice because the τ parameter is harder to tune, as evidenced by the need to assign it different values in the two domains. As Sutton and Barto write:

> Most people find it easier to set the ε parameter with confidence; setting τ requires knowledge of the likely action values and of powers of e (149, pages 27-30).

In this light, interval estimation may be the best choice. Our experiments show that it performs as well or better than softmax and anecdotal evidence suggests that the α parameter is not overly troublesome to tune.

Chapter 4
Evolutionary Function Approximation

The methods presented in Chapter 3 allow the representation-learning capacity of evolutionary algorithms like NEAT to be harnessed in both off-line and on-line scenarios. However, that capacity is still limited in scope to policy search methods. Hence, Sutton and Barto's criticism (that policy search methods, unlike temporal difference methods, do not exploit the specific structure of the reinforcement learning problem) still applies. To address this problem, we need methods that can optimize representations, not just for policies, but value function approximators trained with temporal difference methods.

At present, temporal difference methods typically require a human designer to manually design an appropriate representation for the function approximator. Poor design choices can result in estimates that diverge from the optimal value function (13) and agents that perform poorly. Even for methods with guaranteed convergence (14; 76), achieving high performance in practice requires finding an appropriate representation for the function approximator. As Lagoudakis and Parr observe:

> The crucial factor for a successful approximate algorithm is the choice of the parametric approximation architecture(s) and the choice of the projection (parameter adjustment) method (76, p. 1111).

Nonetheless, representational choices are typically made manually, based only on the designer's intuition.

This chapter introduces *evolutionary function approximation* (161), a new approach to TD function approximation which harnesses the representation-learning power of evolutionary methods. This approach synthesizes evolutionary and TD methods into a single method that automatically selects function approximator representations that enable efficient individual learning. When evolutionary methods are applied to reinforcement learning problems, they typically evolve a population of action selectors, each of which remains fixed during its fitness evaluation. The central insight behind evolutionary function approximation is that, if evolution is directed to evolve value functions instead, then those value functions can be updated, using TD methods, during each fitness evaluation. In this way, the system can *evolve* function approximators that are better able to *learn* via TD. This

S. Whiteson: Adaptive Representations for Reinforcement Learning, SCI 291, pp. 31–46.
springerlink.com

biologically intuitive combination has been applied to computational systems in the past (61; 2; 25; 50; 56; 106) but never, to our knowledge, to aid the discovery of good temporal difference function approximators.

This approach requires only 1) an evolutionary algorithm capable of optimizing representations from a class of functions and 2) a TD method that uses elements of that class for function approximation. This book focuses on performing evolutionary function approximation with neural networks. There are several reasons for this choice. First, they have great experimental value. Nonlinear function approximators are often the most challenging to use; hence, success for evolutionary function approximation with neural networks is good reason to hope for success with linear methods too. Second, neural networks have great potential for function approximation, since they can represent value functions linear methods cannot (given the same basis functions). Finally, employing neural networks is feasible because they have previously succeeded as TD function approximators (38; 151) and sophisticated methods for optimizing their representations (57; 137) already exist.

In addition to automating the search for effective representations, evolutionary function approximation can enable synergistic effects between evolution and learning. How these effects occur depends on which of two possible approaches is employed. The first possibility is a *Lamarckian* approach, in which the changes made by TD during a given generation are written back into the original genomes, which are then used to breed a new population. The second possibility is a *Darwinian* implementation, in which the changes made by TD are discarded and the new population is bred from the original genomes, as they were at birth.

It has long since been determined that biological systems are Darwinian, not Lamarckian. However, it remains unclear which approach is better computationally, despite substantial research (110; 168; 171). The potential advantage of Lamarckian evolution is obvious: it prevents each generation from having to repeat the same learning. However, Darwinian evolution can be advantageous because it enables each generation to reproduce the genomes that led to success in the previous generation, rather than relying on altered versions that may not thrive under continued alteration. Furthermore, in a Darwinian system, the learning conducted by previous generations can be indirectly recorded in a population's genomes via a phenomenon called the *Baldwin Effect* (15), which has been demonstrated in evolutionary computation (61; 2; 25; 9). The Baldwin Effect occurs in two stages. In the first stage, the learning performed by individuals during their lifetimes speeds evolution, because each individual does not have to be exactly right at birth; it need only be in the right neighborhood and learning can adjust it accordingly. In the second stage, those behaviors that were previously learned during individuals' lifetimes become known at birth. This stage occurs because individuals that possess adaptive behaviors at birth have higher overall fitness and are favored by evolution.

Hence, synergistic effects between evolution and learning are possible regardless of which implementation is used. In Section 4.2.4, we compare the two approaches empirically. The following section details NEAT+Q, the implementation of evolutionary function approximation used in our experiments.

4.1 NEAT+Q

All that is required to make NEAT optimize value functions instead of action selectors is a reinterpretation of its output values. The structure of neural network action selectors (one input for each state feature and one output for each action) is already identical to that of Q-learning function approximators. Therefore, if the weights of the networks NEAT evolves are updated during their fitness evaluations using Q-learning and backpropagation, they will effectively evolve value functions instead of action selectors. Hence, the outputs are no longer arbitrary values; they represent the long-term discounted values of the associated state-action pairs and are used, not just to select the most desirable action, but to update the estimates of other state-action pairs.

Algorithm 6 summarizes the resulting NEAT+Q method. Note that this algorithm is identical to Algorithm 2, except for the delineated section containing lines 13–16. Each time the agent takes an action, the network is backpropagated towards

Algorithm 6 NEAT+Q($S, A, c, p, m_n, m_l, g, e, \alpha, \gamma, \lambda, \varepsilon_{td}$)

```
1: // S: set of all states, A: set of all actions, c: output scale, p: population size
2: // m_n: node mutation rate, m_l: link mutation rate, g: number of generations
3: // e: number of episodes per generation, α: learning rate, γ: discount factor
4: // λ: eligibility decay rate, ε_td: exploration rate
5:
6: P[] ← INIT-POPULATION(S, A, p)       // create new population P with random networks
7: for i ← 1 to g do
8:    for j ← 1 to e do
9:       N, s, s' ← P[j % p], null, INIT-STATE(S)                // select next network
10:      repeat
11:         Q[] ← c× EVAL-NET(N, s')        // compute value estimates for current state
12:         ------------------------------------------------------------------
13:         with-prob(ε_td) a' ← RANDOM(A)            // select random exploratory action
14:         else a' ← argmax_k Q[k]                              // or select greedy action
15:         if s ≠ null then
16:            BACKPROP(N, s, a, (r + γ max_k Q[k])/c, α, γ, λ)           // adjust weights
17:         ------------------------------------------------------------------
18:         s, a ← s', a'
19:         r, s' ← TAKE-ACTION(a')           // take action and transition to new state
20:         N.fitness ← N.fitness + r              // update total reward accrued by N
21:      until TERMINAL-STATE?(s)
22:      N.episodes ← N.episodes + 1         // update total number of episodes for N
23:   P'[] ← new array of size p               // new array will store next generation
24:   for j ← 1 to p do
25:      P'[j] ← BREED-NET(P[])          // make a new network based on fit parents in P
26:      with-probability m_n: ADD-NODE-MUTATION(P'[j]) // add node to new network
27:      with-probability m_l: ADD-LINK-MUTATION(P'[j])        // add link to new network
28:   P[] ← P'[]
```

Q-learning targets (line 16) and ϵ-greedy selection occurs just as in Algorithm 1 (lines 13–14). If α and ϵ_{td} are set to zero, this method degenerates to regular NEAT.

NEAT+Q combines the power of TD methods with the ability of NEAT to learn effective representations. Traditional neural network function approximators put all their eggs in one basket by relying on a single manually designed network to represent the value function. NEAT+Q, by contrast, explores the space of such networks to increase the chance of finding a representation that will perform well.

In NEAT+Q, the weight changes caused by backpropagation accumulate in the current population's networks throughout each generation. When a network is selected for an episode, its weights begin exactly as they were at the end of its last episode. In the Lamarckian approach, those changes are copied back into the networks' genomes and inherited by their offspring. In the Darwinian approach, those changes are discarded at the end of each generation.

4.2 Results

We conducted a series of experiments in the mountain car and server job scheduling domains (described in Section 3.4) to empirically evaluate the methods presented in this chapter. Section 4.2.1 compares manual and evolutionary function approximators. Section 4.2.2 tests evolutionary function approximation combined with on-line evolutionary computation. Section 4.2.3 compares these novel approaches to previous learning and non-learning methods. Section 4.2.4 compares Darwinian and Lamarckian versions of evolutionary function approximation. Finally, Section 4.2.5 presents some additional tests that measure the effect of continual learning on function approximators. The results offer insight into why certain methods outperform others in these domains and what factors can make neural network function approximation difficult in practice.

Each of the graphs presented in these sections include error bars indicating 95% confidence intervals. In addition, to assess statistical significance, we conducted Student's t-tests on each pair of methods evaluated. The results of these tests are summarized in Appendix A.

4.2.1 Comparing Manual and Evolutionary Function Approximation

For an initial baseline, we used the same off-line NEAT results presented in Section 3.5. Next, we performed 25 runs in each domain using NEAT+Q, with the same parameter settings. The eligibility decay rate λ was 0.0. and the learning rate α was set to 0.1 and annealed linearly for each member of the population until reaching zero after 100 episodes.[1] In scheduling, γ was 0.95 and ϵ_{td} was 0.05. Those values of γ and ϵ_{td} work well in mountain car too, though in the experiments

[1] Other values of λ were tested in the context of NEAT+Q but had little effect on performance.

presented here they were set to 1.0 and 0.0 respectively, since Sutton (144) found that discounting and exploration are unnecessary in mountain car. The output scale c was set to -100 in mountain car and -1000 in scheduling.

We tested both Darwinian and Lamarckian NEAT+Q in this manner. Both perform well, though which is preferable appears to be domain dependent. For simplicity, in this section and those that follow, we present results only for Darwinian NEAT+Q. In Section 4.2.4 we present a comparison of the two approaches.

To test Q-learning without NEAT, we tried 24 different configurations in each domain. These configurations correspond to every possible combination of the following parameter settings. The networks had feed-forward topologies with 0, 4, or 8 hidden nodes. The learning rate α was either 0.01 or 0.001. The annealing schedules for α were linear, decaying to zero after either 100,000 or 250,000 episodes. The eligibility decay rate λ was either 0.0 or 0.6. The other parameters, γ and ε, were set just as with NEAT+Q, and the standard deviation of initial weights σ was 0.1. Each of these 24 configurations was evaluated for 5 runs. In addition, we experimented informally with higher and lower values of α, higher values of γ, slower linear annealing, exponential annealing, and no annealing at all, though none performed as well as the results presented here.

In these experiments, each run used a different set of initial weights. Hence, the resulting performance of each configuration, by averaging over different initial weight settings, does not account for the possibility that some weight settings perform consistently better than others. To address this, for each domain, we took the best performing configuration[2] and randomly selected five fixed initial weight settings. For each setting, we conducted 5 additional runs. Finally, we took the setting with the highest performance and conducted an additional 20 runs, for a total of 25. For simplicity, the graphs that follow show only this Q-learning result: the best configuration with the best initial weight setting.

Figure 4.1 shows the results of these experiments. For each method, the corresponding line in the graph represents a uniform moving average over the aggregate reward received in the past 1,000 episodes, averaged over all 25 runs. Using average performance, as we do throughout this book, is somewhat unorthodox for evolutionary methods, which are more commonly evaluated on the performance of the generation champion. There are two reasons why we adopt average performance. First, it creates a consistent metric for all the methods tested, including the TD methods that do not use evolutionary computation and hence have no generation champions. Second, it is an on-line metric because it incorporates *all* the reward the learning system accrues. Plotting only generation champions is an implicitly off-line metric because it does not penalize methods that discover good policies but fail to accrue much reward while learning. Hence, average reward is a better metric for evaluating on-line evolutionary computation, as we do in Section 4.2.2.

To make a larger number of runs computationally feasible, both NEAT and NEAT+Q were run for only 100 generations. In the scheduling domain, neither

---

[2] Mountain car parameters were: 4 hidden nodes, $\alpha = 0.001$, annealed to zero at episode 100,000, $\lambda = 0.0$. Server job scheduling parameters were: 4 hidden nodes, $\alpha = 0.01$, annealed to zero at episode 100,000, $\lambda = 0.6$.

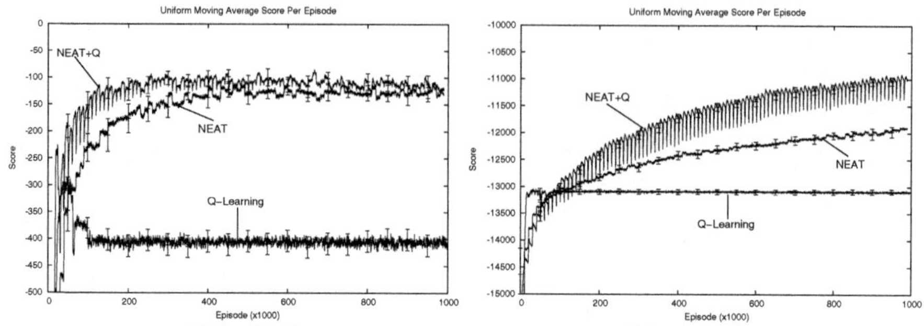

(a) Mountain Car (b) Server Job Scheduling

Fig. 4.1 A comparison of the performance of NEAT, NEAT+Q, and Q-learning with the best of 24 different manually designed neural network function approximators in the mountain car and server job scheduling domains.

method has completely plateaued by this point. However, a handful of trials conducted for 200 generations verified that only very small additional improvements are made after 100 generations, without a qualitative effect on the results.

Note that the progress of NEAT+Q consists of a series of 10,000-episode intervals. Each of these intervals corresponds to one generation and the changes within them are due to learning via Q-learning and backpropagation. Although each individual learns for only 100 episodes on average, NEAT's system of randomly selecting individuals for evaluation causes that learning to be spread across the entire generation: each individual changes gradually during the generation as it is repeatedly evaluated. The result is a series of intra-generational learning curves within the larger learning curve.

For the particular problems we tested and network configurations we tried, evolutionary function approximation significantly improves performance over manually designed networks. In the scheduling domain, Q-learning learns much more rapidly in the very early part of learning. In both domains, however, Q-learning soon plateaus while NEAT and NEAT+Q continue to improve. Of course, after 100,000 episodes, Q-learning's learning rate α has annealed to zero and no additional learning is possible. However, its performance plateaus well before α reaches zero and, in our experiments, running Q-learning with slower annealing or no annealing at all consistently led to inferior and unstable performance.

Nonetheless, the possibility remains that additional engineering of the network structure, the feature set, or the learning parameters would significantly improve Q-learning's performance. In particular, when Q-learning is started with one of the best networks discovered by NEAT+Q and the learning rate is annealed aggressively, Q-learning matches NEAT+Q's performance without directly using evolutionary computation. However, it is unlikely that a manual search, no matter how extensive, would discover these successful topologies, which contain irregular and partially connected hidden layers. Figure 4.2 shows examples of typical networks evolved by NEAT+Q.

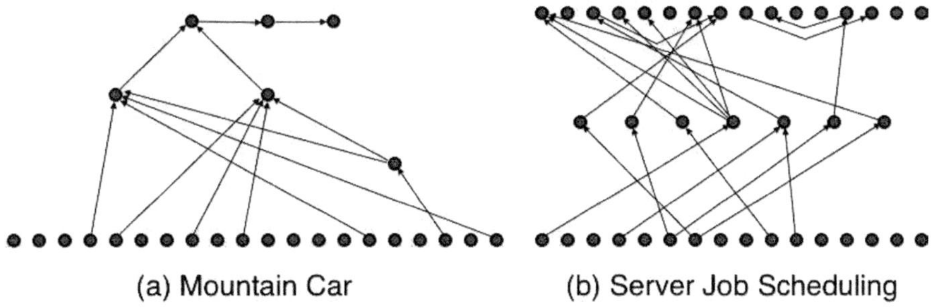

Fig. 4.2 Typical examples of the topologies of the best networks evolved by NEAT+Q in both the mountain car and scheduling domains. Input nodes are on the bottom, hidden nodes in the middle, and output nodes on top. In addition to the links shown, each input node is directly connected to each output node. Note that two output nodes can be directly connected, in which case the activation of one node serves not only as an output of the network, but as an input to the other node.

NEAT+Q also significantly outperforms regular NEAT in both domains. In the mountain car domain, NEAT+Q learns faster, achieving better performance in earlier generations, though both plateau at approximately the same level. In the server job scheduling domain, NEAT+Q learns more rapidly and also converges to significantly higher performance. This result highlights the value of TD methods on challenging reinforcement learning problems. Even when NEAT is employed to find effective representations, the best performance is achieved only when TD methods are used to estimate a value function. Hence, the relatively poor performance of Q-learning is not due to some weakness in the TD methodology but merely to the failure to find a good representation.

Furthermore, in the scheduling domain, the advantage of NEAT+Q over NEAT is not directly explained just by the learning that occurs via backpropagation within each generation. After 300,000 episodes, NEAT+Q clearly performs better even at the beginning of each generation, before such learning has occurred. Just as predicted by the Baldwin Effect, evolution proceeds more quickly in NEAT+Q because the weight changes made by backpropagation, in addition to improving that individual's performance, alter selective pressures and more rapidly guide evolution to useful regions of the search space.

4.2.2 Combining On-Line Evolution with Evolutionary Function Approximation

Sections 3.5 and 4.2.1 verify that both on-line evolutionary computation and evolutionary function approximation can significantly boost performance in reinforcement learning tasks. This section presents experiments that assess how well these two ideas work together.

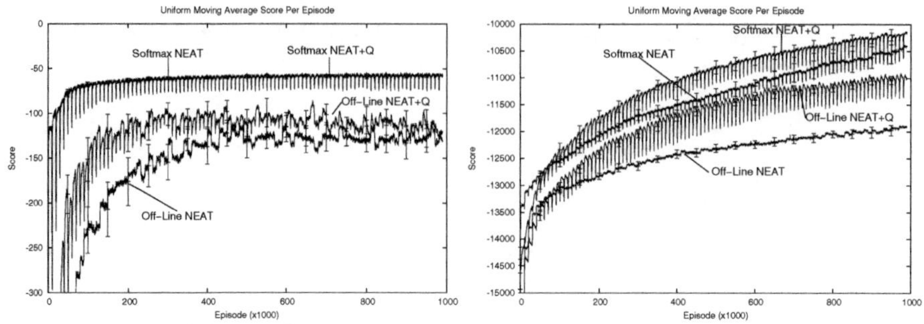

(a) Mountain Car (b) Server Job Scheduling

Fig. 4.3 The performance of combining evolutionary function approximation with on-line evolutionary computation compared to using each individually in the mountain car and server job scheduling domains

Figure 4.3 presents the results of combining NEAT+Q with softmax evolutionary computation, averaged over 25 runs, and compares it to using each of these methods individually, i.e., using off-line NEAT+Q (as done in Section 4.2.1) and using softmax evolutionary computation with regular NEAT. For simplicity, we do not present results for ϵ-greedy or interval estimation NEAT+Q since softmax NEAT+Q performed the best in Section 3.5.

In both domains, softmax NEAT+Q performs significantly better than off-line NEAT+Q. Hence, just like regular evolutionary computation, evolutionary function approximation performs better when supplemented with selection techniques traditionally used in TD methods. Surprisingly, in the mountain car domain, softmax NEAT+Q performs only as well softmax NEAT. We attribute these results to a ceiling effect, i.e., the mountain car domain is easy enough that, given an appropriate selection mechanism, NEAT is able to learn quite rapidly, even without the help of Q-learning. In the server job scheduling domain, softmax NEAT+Q does perform better than softmax NEAT, though the difference is rather modest. Hence, in both domains, the most critical factor to boosting the performance of evolutionary computation is the use of an appropriate selection mechanism.

4.2.3 *Comparing to Other Approaches*

The experiments presented thus far verify that the novel methods presented in this chapter can improve performance over the constituent techniques upon which they are built. This section presents experiments that compare the performance of the highest performing novel method, softmax NEAT+Q, to previous approaches. In the mountain car domain, we compare to previous results that use TD methods with a linear function approximator (144). In the server job scheduling domain, we compare to a random scheduler, two non-learning schedulers from previous research (95; 160), and an analytical solution computed using integer linear programming.

In the mountain car domain, the results presented above make clear that softmax NEAT+Q can rapidly learn a good policy. However, since these results use an on-line metric, performance is averaged over all members of the population. Hence, they do not reveal how close the best learned policies are to optimal. To assess the best policies, we selected the generation champion from the final generation of each softmax NEAT+Q run and evaluated it for an additional 1,000 episodes. Then we compared the results to the performance of a learner using Sarsa, a TD method similar to Q-learning (149), with tile coding, a popular linear function approximator (4), using a setup that matches that of Sutton (144) as closely as possible. We found their performance to be nearly identical: softmax NEAT+Q received an average score of -52.75 while the Sarsa tile coding learner received -52.02. We believe this performance is approximately optimal, as it matches the best results published by other researchers, e.g. (129).

This result does not imply that neural networks are the function approximator of choice for the mountain car domain. On the contrary, Sutton's tile coding converges in many fewer episodes. Nonetheless, these results demonstrate that evolutionary function approximation and on-line evolution make it feasible to find approximately optimal policies using neural networks, something that some previous approaches (28; 115), using manually designed networks, were unable to do.

Since the mountain car domain has only two state features, it is possible to visualize the value function. Figure 4.4 compares the value functions learned by softmax NEAT+Q to that of Sarsa with tile coding. For clarity, the graphs plot estimated steps to the goal. Since the agent receives a reward of -1 for each timestep until reaching the goal, this quantity is equivalent to $-\max_a(Q(s,a))$. Surprisingly, the two value functions bear little resemblance to one another. While they share some very general characteristics, they differ markedly in both shape and scale. Hence, these graphs highlight a fact that has been noted before (151): that TD methods can learn excellent policies even if they estimate the value function only very grossly. So long as the value function assigns the highest value to the correct action, the agent will perform well.

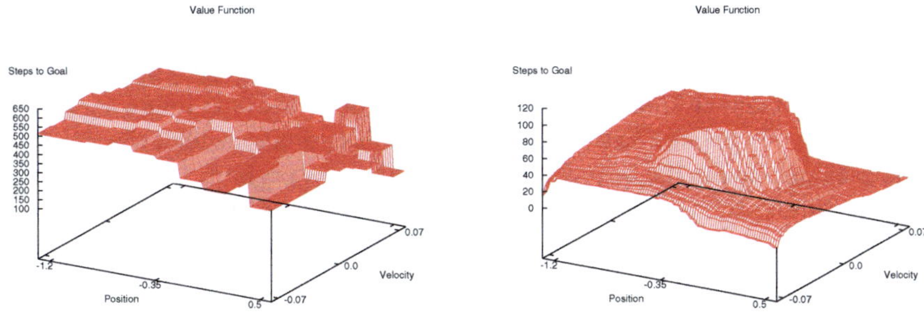

(a) NEAT+Q Network (b) CMAC

Fig. 4.4 The value function, shown as estimated steps to the goal, of policies learned by softmax NEAT+Q and Sarsa using tile coding

In the server job scheduling domain, finding alternative approaches for comparison is less straightforward. Substantial research about job scheduling already exists but most of the methods involved are not applicable here because they do not allow jobs to be associated with arbitrary utility functions. For example, Liu and Layland (86) present methods for job scheduling in a real-time environment, in which a hard deadline is associated with each job. McWherter et al. (94) present methods for scheduling jobs with different priority classes. However, unlike the utility functions shown in Section 3.4.2, the relative importance of a job type does not change as a function of time. McGovern et al. (91) use reinforcement learning for CPU instruction scheduling but aim only to minimize completion time.

One method that can be adapted to the server job scheduling task is the generalized $c\mu$ rule (95), in which the server always processes at time t the oldest job of that type k which maximizes $C'_k(o_k)/p_k$, where C'_k is the derivative of the cost function for job type k, o_k is the age of the oldest job of type k and p_k is the average processing time for jobs of type k. Since in our simulation all jobs require unit time to process and the cost function is just the additive inverse of the utility function, this algorithm is equivalent to processing the oldest job of that type k that maximizes $-U'_k(o_k)$, where U'_k is the derivative of the utility function for job type k. The generalized $c\mu$ rule has been proven approximately optimal given convex cost functions (95). Since the utility functions, and hence the cost functions, are both convex and concave in our simulation, there is no theoretical guarantee about its performance in the server job scheduling domain. To see how well it performs in practice, we implemented it in our simulator and ran it for 1,000 episodes, obtaining an average score of $-10{,}891$.

Another scheduling algorithm applicable to this domain is the insertion scheduler, which performed the best in a previous study of a very similar domain (160). The insertion scheduler uses a simple, fast heuristic: it always selects for processing the job at the head of the queue but it keeps the queue ordered in a way it hopes will maximize aggregate utility. For any given ordering of a set of J jobs, the aggregate utility is:

$$\sum_{i \in J} U_i(a_i + p_i)$$

where $U_i(\cdot)$, a_i, and p_i are the utility function, current age, and position in the queue, respectively, of job i. Since there are $|J|!$ ways to order the queue, it is clearly infeasible to try them all. Instead, the insertion scheduler uses the following simple, fast heuristic: every time a new job is created, the insertion scheduler tries inserting it into each position in the queue, settling on whichever position yields the highest aggregate utility. Hence, by bootstrapping off the previous ordering, the insertion scheduler must consider only $|J|$ orderings. We implemented the insertion scheduler in our simulator and ran it for 1,000 episodes, obtaining an average score of $-13{,}607$.

Neither the $c\mu$ rule nor the insertion scheduler perform as well as softmax NEAT+Q, whose final generation champions received an average score of $-9{,}723$ over 1,000 episodes. Softmax NEAT+Q performed better despite the fact that the alternatives rely on much greater *a priori* knowledge about the dynamics of the system. Both alternatives require the scheduler to have a predictive model of the system, since their calculations depend on knowledge of the utility functions and the amount of time each job takes to complete. By contrast, softmax NEAT+Q, like many reinforcement learning algorithms, assumes such information is hidden and discovers a good policy from experience, just by observing state transitions and rewards.

If, in addition to assuming the scheduler has a model of the system, we make the unrealistic assumption that unlimited computation is available to the scheduler, then we can obtain an informative upper bound on performance. At each time step of the simulation, we can compute the optimal action analytically by treating the scheduling problem as an integer linear program. For each job $i \in J$ and for each position j in which it could be placed, the linear program contains a variable $x_{ij} \in \{0,1\}$. Associated with each variable is a weight $w_{ij} = U_i(a_i + j)$, which represents the reward the scheduler will receive when job i completes given that it currently resides in position j. Since the scheduler's goal is to maximize aggregate utility, the linear program must maximize $\sum_i \sum_j w_{ij} x_{ij}$. In addition to the constraint that $\forall ij : x_{ij} \in \{0,1\}$, the program is also constrained such that each job is in exactly one position: $\forall i : \sum_j x_{ij} = 1$ and that each position holds exactly one job: $\forall j : \sum_i x_{ij} = 1$.

A solution to the resulting integer linear program is an ordering that will maximize the aggregate utility of the jobs currently in the queue. If the scheduler always processes the job in the first position of this ordering, it will behave optimally *assuming no more jobs arrive*. Since new jobs are constantly arriving, the linear program must be re-solved anew at each time step. The resulting behavior may still be suboptimal since the decision about which job to process is made without reasoning about what types of jobs are likely to arrive later. Nonetheless, this analytical solution represents an approximate upper bound on performance in this domain.

Using the CPLEX software package, we implemented a scheduler based on the linear program described above and tested in our simulator for 1,000 episodes, obtaining an average score of −7,819. Not surprisingly, this performance is superior to that of softmax NEAT+Q, though it takes, on average, 741 times as long to run. The computational requirements of this solution are not likely to scale well either, since the number of variables in the linear program grows quadratically with respect to the size of the queue.

Figure 4.5 summarizes the performance of the alternative scheduling methods described in this section and compares them to softmax NEAT+Q. It also includes, as a lower bound on performance, a random scheduler, which received an average score of −15,502 over 1,000 episodes. A Student's t-test verified that the difference in performance between each pair of methods is statistically significant with 95% confidence. Softmax NEAT+Q performs the best except for the linear programming approach, which is computationally expensive and relies on a model of the system. Prior to learning, softmax NEAT+Q performs similarly to the random scheduler. The difference in performance between the best learned policies and the linear programming upper bound is 75% better than that of the baseline random scheduler and 38% better than that of the next best method, the $c\mu$ scheduler.

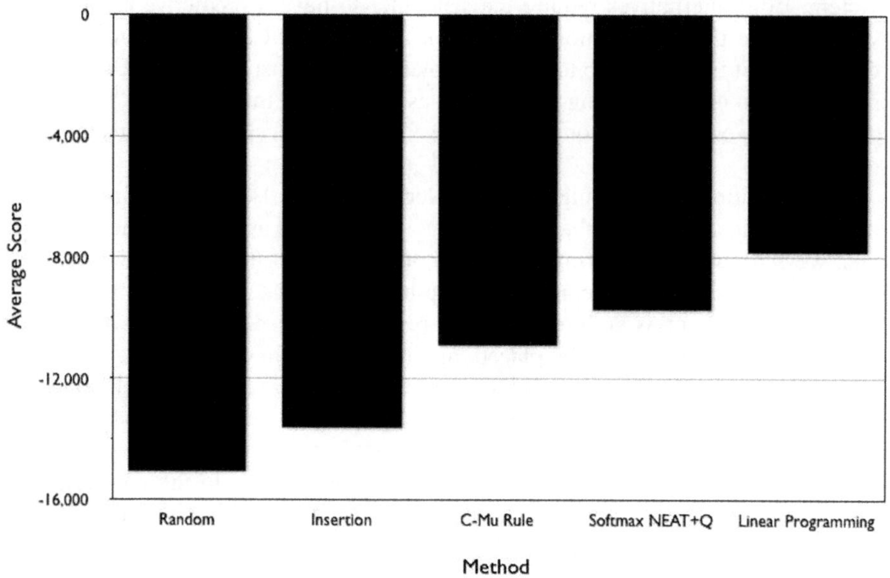

Fig. 4.5 A comparison of the performance of softmax NEAT+Q and several alternative methods in the server job scheduling domain

4.2.4 Comparing Darwinian and Lamarckian Approaches

As described in the beginning of this chapter, evolutionary function approximation can be implemented in either a Darwinian or Lamarckian fashion. The results presented so far all use the Darwinian implementation of NEAT+Q. However, it is not clear that this approach is superior even though it more closely matches biological systems. In this section, we compare the two approaches empirically in both the mountain car and server job scheduling domains. Many other empirical comparisons of Darwinian and Lamarckian systems have been conducted previously (168; 171; 110) but ours is novel in that individual learning is based on a TD function approximator. In other words, these experiments address the question: when trying to approximate a TD value function, is a Darwinian or Lamarckian approach superior?

Figure 4.6 compares the performance of Darwinian and Lamarckian NEAT+Q in both the mountain car and server job scheduling domains. In both cases, we use off-line NEAT+Q, as the on-line versions tend to mute the differences between the two implementations. Though both implementations perform well in both domains, Lamarckian NEAT+Q does better in mountain car but worse in server job scheduling. Hence, the relative performance of these two approaches seems to depend critically on the dynamics of the domain to which they are applied. In the following

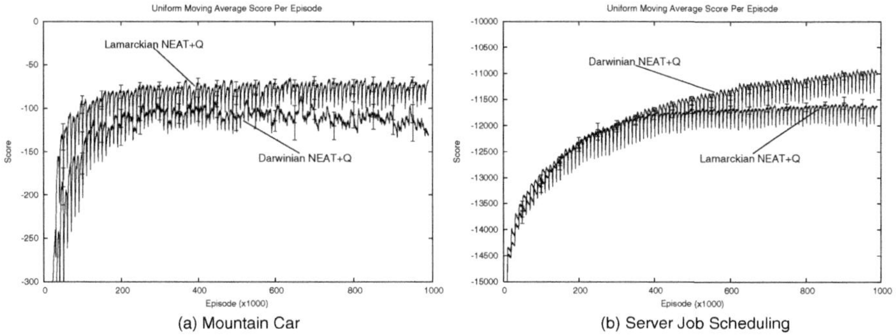

Fig. 4.6 A comparison of Darwinian and Lamarckian NEAT+Q in the mountain car and server job scheduling domains

section, we present some additional results that elucidate which factors affect their performance.

4.2.5 Continual Learning Tests

In this section, we assess the performance of the best networks discovered by NEAT+Q when evaluated for many additional episodes. We compare two scenarios, one where the learning rate is annealed to zero after 100 episodes, just as in training, and one where it is not annealed at all. Comparing performance in these two scenarios allows us to assess the effect of continual learning on the evolved networks.

We hypothesized that NEAT+Q's best networks would perform well under continual learning in the mountain car domain but not in server job scheduling. This hypothesis was motivated by the results of early experiments with NEAT+Q. Originally, we did not anneal α at all. This setup worked fine in the mountain car domain but in scheduling it worked only with off-line NEAT+Q; on-line NEAT+Q actually performed worse than off-line NEAT+Q! Annealing NEAT+Q's learning rate eliminated the problem, as the experiments in Section 4.2.2 verify. If finding weights that remain stable under continual learning is more difficult in scheduling than in mountain car, it could explain this phenomenon, since ϵ-greedy and softmax selection, by giving many more episodes of learning to certain networks, could cause those networks to become unstable and perform poorly.

To test the best networks without continual learning, we selected the final generation champion from each run of off-line Darwinian NEAT+Q and evaluated it for an additional 5,000 episodes, i.e., 50 times as many episodes as it saw in training. During these additional episodes, the learning rate was annealed to zero by episode 100, just as in training. To test the best networks with continual learning, we repeated this experiment but did not anneal the learning rate at all. To prevent any unnecessary discrepancies between training and testing, we repeated the original NEAT+Q runs with annealing turned off and used the resulting final generation champions.

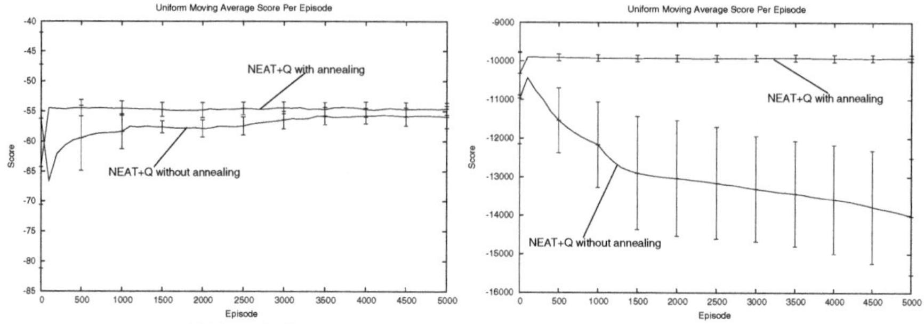

(a) Mountain Car (b) Server Job Scheduling

Fig. 4.7 A comparison of the performance of the best networks evolved by NEAT+Q when tested, with and without annealing, for an additional 5,000 episodes

Figure 4.7 shows the results of these tests. In the mountain car domain, performance remains relatively stable regardless of whether the networks continue to learn. The networks tested without annealing show more fluctuation but maintain performance similar to those that were annealed. However, in the scheduling domain, the networks subjected to continual learning rapidly plummet in performance whereas those that are annealed continue to perform as they did in training. These results directly confirm our hypothesis that evolutionary computation can find weights that perform well under continual learning in mountain car but not in scheduling, which explains why on-line NEAT+Q does not require an annealed learning rate in mountain car but does in scheduling.

These tests also shed light on the comparison between Darwinian and Lamarckian NEAT+Q presented in Section 4.2.4. A surprising feature of the Darwinian approach is that it is insensitive to the issue of continual learning. Since weight changes do not affect offspring, evolution need only find weights that remain suitable during one individual's lifetime. By contrast, in the Lamarckian approach, weight changes accumulate from generation to generation. Hence, the TD updates that helped in early episodes can hurt later on. In this light it makes perfect sense that Lamarckian NEAT+Q performs better in mountain car than in scheduling, where continual learning is problematic.

These results suggest that the problem of stability under continual learning can greatly exacerbate the difficulty of performing neural network function approximation in practice. This issue is not specific to NEAT+Q, since Q-learning with manually designed networks achieved decent performance only when the learning rate was properly annealed. Darwinian NEAT+Q is a novel way of coping with this problem, since it obviates the need for long-term stability. In on-line evolutionary computation annealing may still be necessary but it is less critical to set the rate of decay precisely. When learning ends, it prevents only a given individual from continuing to improve. The system as a whole can still progress, as evolution exerts selective pressure and learning begins anew in the next generation.

4.3 Discussion

The results in the mountain car domain presented in this chapter demonstrate that NEAT+Q can successfully train neural network function approximators in a domain which is notoriously problematic for them. However, NEAT+Q requires many more episodes to find good solutions (by several orders of magnitude) than tile coding does in the same domain. This contrast highlights an important drawback of NEAT+Q: since each candidate network must be trained long enough to let Q-learning work, it has very high sample complexity. However, in the next chapter, we introduce an enhancement to NEAT+Q that dramatically reduces its sample complexity.

It is not surprising that NEAT+Q takes longer to learn than tile coding because it is actually solving a more challenging problem. Tile coding, like other linear function approximators, requires the human designer to engineer a state representation in which the optimal value function is linear with respect to those state features (or can be reasonably approximated as such). For example, when tile coding was applied to the mountain car domain, the two state features were tiled conjunctively (144). By contrast, nonlinear function approximators like neural networks can take a simpler state representation and *learn* the important nonlinear relationships. Note that the state representation used by NEAT+Q, while discretized, does not include any conjunctive features of the original two state features. The important conjunctive features are represented by hidden nodes that are evolved automatically by NEAT.

Conjunctively tiling all state features is feasible in mountain car but quickly becomes impractical in domains with more state features. For example, doing so in the scheduling domain would require 16 tile codings, one for each action. In addition, each tile coding would have multiple 16-dimensional tilings. If 10 tilings were used and each state feature were discretized into 10 buckets, the resulting function approximator would have $16 \times 10 \times 10^{16}$ cells. Conjunctively tiling only some state features is feasible only with a large amount of domain expertise. Hence, methods like NEAT+Q that automatically learn nonlinear representations promise to be of great practical importance.

The results in the scheduling domain demonstrate that the proposed methods scale to a much larger, probabilistic domain and can learn schedulers that outperform existing non-learning approaches. The difference in performance between the best learned policies and the linear programming upper bound is 75% better than that of the baseline random scheduler and 38% better than that of the next best method, the $c\mu$ scheduler. However, the results also demonstrate that non-learning methods can do quite well in this domain. If so, is it worth the trouble of learning? We believe so. In a real system, the utility functions that the learner maximizes would likely be drawn directly from Service Level Agreements (SLAs), which are legally binding contracts governing how much clients pay their service providers as a function of the quality of service they receive (157). Hence, even small improvements in system performance can significantly affect the service provider's bottom line. Substantial improvements like those demonstrated in our results, if replicated in real systems, could be very valuable indeed.

Overall, the main limitation of the results presented in this chapter is that they apply only to neural networks. In particular, the analysis about the effects of continual learning (Section 4.2.5) may not generalize to other types of function approximation that are not as prone to instability or divergence if over-trained. While evolutionary methods could in principle be combined with any kind of function approximation, in practice it is likely to work well only with very concise representations. Methods like tile coding, which use many more weights, would result in very large genomes and hence be difficult for evolutionary computation to optimize. However, as Chapter 7 will demonstrate, other strategies which do not rely on evolutionary computation can effectively optimize such representations.

Chapter 5
Sample-Efficient Evolutionary Function Approximation

As mentioned in Section 4.3, evolutionary function approximation suffers from one important disadvantage: high sample complexity. Each candidate representation in the population must be evaluated for many episodes before TD updates have a significant effect. High sample complexity is undesirable because sample episodes are typically the scarcest resource: each new episode may incur substantial real-world costs whereas additional memory and CPU cycles are relatively inexpensive.

This chapter presents an enhancement to evolutionary function approximation designed to make it dramatically more sample-efficient. This enhancement relies on TD methods that are *off-policy*, i.e., that can estimate the optimal value function regardless of what policy the agent is following. By storing experience from the previous generation, sample-efficient evolutionary function approximation can train each new generation off-line using only computation time: no additional sample episodes are needed. The resulting function approximators can then be evaluated and selectively reproduced in many fewer episodes.

We implemented this enhancement in NEAT and tested the resulting sample-efficient NEAT+Q algorithm in a deterministic variant of server job scheduling. The results demonstrate that sample-efficient NEAT+Q can learn better policies than NEAT or Q-learning alone and can do so in many fewer episodes than the original NEAT+Q approach.

5.1 Sample-Efficient NEAT+Q

For both NEAT and NEAT+Q, the number of episodes per generation e must be much greater than the population size $|P|$ in domains that are highly stochastic. Such domains have noisy fitness functions and hence each network's performance must be averaged over many episodes. For NEAT+Q, however, there is a second reason to set e high, which applies even if the domain is deterministic: Q-learning needs time to learn. In most domains, TD updates will not have substantial impact in a single episode. Consequently, the original NEAT+Q method is likely to offer a practical

S. Whiteson: Adaptive Representations for Reinforcement Learning, SCI 291, pp. 47–52.
springerlink.com

advantage over regular NEAT only in highly stochastic domains, where e must be set high anyway. Otherwise, even if NEAT+Q ultimately discovers better policies, it will take many more episodes to do so. Figure 5.1 illustrates this problem.

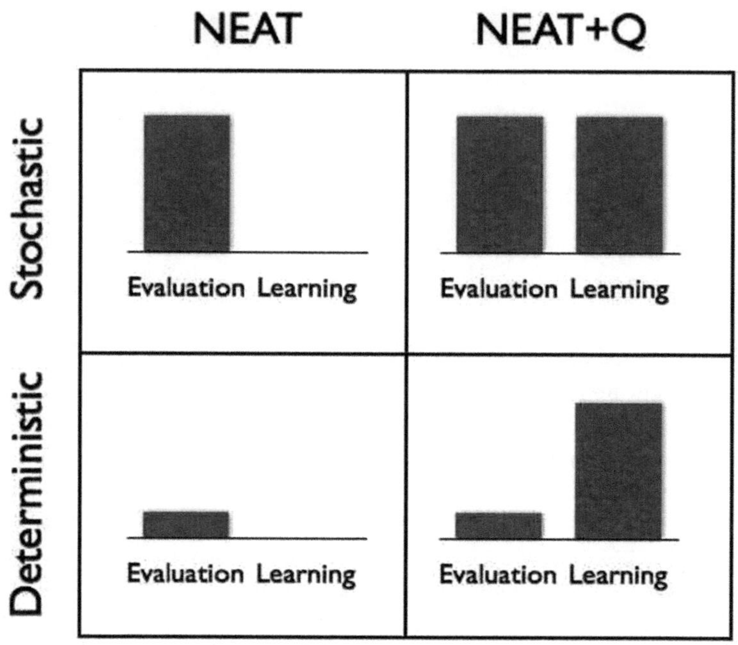

Fig. 5.1 A comparison of the number of episodes necessary for evaluation and learning in both NEAT and NEAT+Q. No learning occurs in regular NEAT and evaluations can occur in a single episode in deterministic domains (bottom left) but require many episodes in stochastic domains (top left). NEAT+Q requires many episodes to train each network but in stochastic domains, those episodes were already necessary for evaluation (top right). Hence, the problematic case for NEAT+Q occurs in deterministic domains, where many more episodes are required for learning than would have been for evaluation (bottom right).

This section presents sample-efficient NEAT+Q (163), a variation designed to remedy this shortcoming. By training networks on saved experience, Q-learning can have a substantial impact even when $e = |P|$. As a result, NEAT+Q can improve performance even in completely deterministic domains. Sample-efficient NEAT+Q works by exploiting the off-policy nature of Q-learning. Because Q-learning's update rule is independent of the policy the agent is following, one network can be updated while another is controlling the agent. Furthermore, a network can be updated based on data saved from previous sample episodes, regardless of what policy was used during those episodes. Consequently, it is not necessary to use different episodes to train each network. On the contrary, by saving data from the episodes used by the previous generation, each network in the population can be pre-trained, using computation time but no additional sample episodes. If the fitness function is not too noisy then, once trained, the resulting function approximators can be evaluated by NEAT+Q using only $|P|$ episodes.

To achieve this sample-efficiency, NEAT+Q records all the transition samples, of the form (s, a, r, s'), from the episodes used to evaluate the previous generation. Then, at the beginning of each generation (i.e., after line 7 in Algorithm 6), it calls the PRE-TRAIN function described in Algorithm 7. In the first generation no samples have been collected ($|T| = 0$) and no pre-training occurs.

Algorithm 7 PRE-TRAIN($P, T, c, \alpha, \gamma, \lambda$)

```
// P: population, T: sample transitions, c: output scale, α: learning rate
// γ: discount factor, λ: eligibility decay rate

for i ← 1 to |P| do
    for j ← 1 to |T| do
        Q[] ← EVAL-NET(P[i], T[j].s')
        BACKPROP(P[i], T[j].s, T[j].a, T[j].r + γ max_k Q[k]/c, α, γ, λ)
```

Because it saves sample episodes for reuse, sample-efficient NEAT+Q bears a close resemblance to experience replay methods for reinforcement learning (84). In particular, it is similar to Neural Fitted Q Iteration (119), which uses data from saved episodes to train neural network TD function approximators. The primary difference is that these methods do not learn representations because they use saved experience to train only one function approximator. By contrast, sample-efficient NEAT+Q uses saved experience to train an entire population of function approximators with heterogeneous representations and then subjects them to evolutionary selection.

If computational resources are plentiful, there are many ways to extend the pre-training phase. For example, episodes could be saved from all previous generations instead of just the last one and/or each network could be trained repeatedly on each sample instead of just once. To make our experiments more feasible, we do not evaluate these alternatives in this chapter. However, the experiments presented below suggest that additional pre-training does not improve performance.

Assuming e is reduced to $|P|$, this algorithm will have much higher amortized computational complexity per episode than the original NEAT+Q method, since each network must be trained before evaluations can begin. However, it will have much lower sample complexity since each generation requires many fewer episodes. This trade-off is likely to be advantageous in practice, since sample experience is typically a much scarcer resource than computation time.

5.2 Results

In this chapter, we consider a deterministic variation of the server job scheduling task that was introduced in Section 3.4.2. At the beginning of each learning run, we randomly select the sequence of 200 jobs that the agent will process in each episode. Hence, within each run, every episode uses the same sequence of jobs, though that sequence differs for each run. Making the task deterministic allows us to evaluate sample-efficient NEAT+Q in the scenario for which it was designed. In the stochastic version of the task, the fitness function is very noisy and each network must be evaluated for approximately 100 episodes to get an accurate fitness estimate, giving the original NEAT+Q method enough time to significantly improve performance. In the deterministic version, each network can be accurately evaluated in a single episode and hence NEAT+Q will significantly improve performance only if it is made sample-efficient.

Figure 5.2 shows the performance of NEAT and NEAT+Q in the deterministic scheduling task, with $|P| = e = 50$. The graph shows uniform moving average score per episode averaged over the past 100 episodes. The performance advantage of NEAT+Q that was shown in Section 4.2.1 disappears because Q-learning does not have a significant effect in one episode.

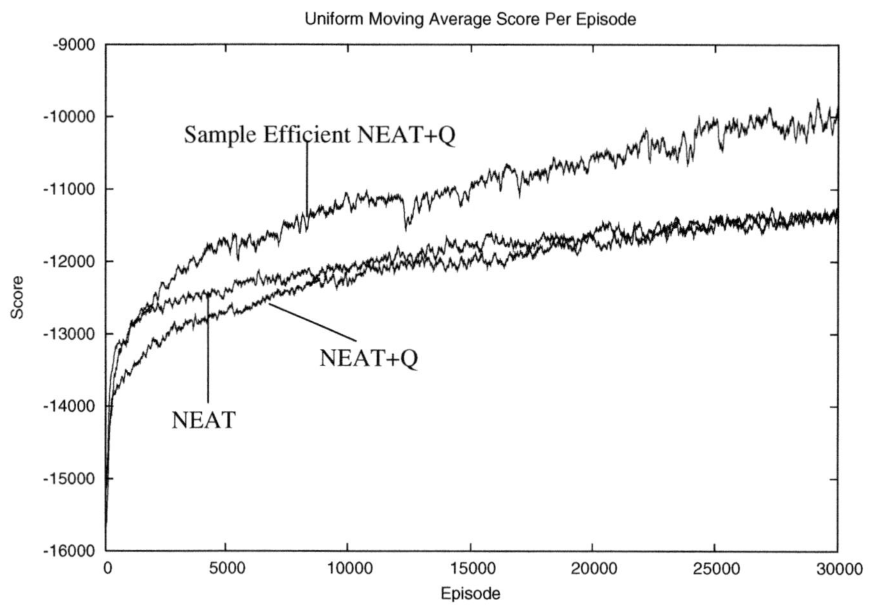

Fig. 5.2 A comparison of the performance of NEAT with both regular and sample-efficient versions of NEAT+Q in the deterministic server job scheduling task

Nonetheless, NEAT+Q can substantially improve performance even in the deterministic version of the task if it is made sample-efficient. Figure 5.2 also shows the performance of sample-efficient NEAT+Q. By pre-training with saved episodes, this method substantially outperforms regular NEAT and the original NEAT+Q method.

Obtaining this performance improvement requires additional computation time (pre-training requires 10,000 neural network updates for each member of the population) but the result is a dramatic drop in sample complexity. By saving episodes, NEAT+Q can outperform NEAT even when the number of episodes per generation is reduced by two orders of magnitude. A Student's t-test confirmed that the performance difference between sample-efficient NEAT+Q and both regular NEAT+Q and NEAT is statistically significant with 95% confidence.

In these experiments, sample-efficient NEAT+Q saves transitions from all the episodes used to evaluate the previous generation. Hence, each network is pre-trained with 50 sample episodes. Would performance improve more if additional episodes were saved? Could the same performance be achieved with less computation time if fewer episodes were saved? To address these questions, we ran additional trials of sample-efficient NEAT+Q, pre-training on 5, 10, 25, or 100 episodes instead of 50. Figure 5.3 summarizes the results of these experiments by comparing the average performance of each method after 30,000 episodes. Surprisingly, pre-training with as few as 5 saved episodes (1,000 updates per network) still yields a substantial performance advantage. Furthermore, pre-training with 100 episodes (20,000 updates per network) does not improve performance. A Student's t-test demonstrated that, while the differences between each sample-efficient version of NEAT+Q and both regular NEAT+Q and NEAT are significant, the differences among them are not significant.

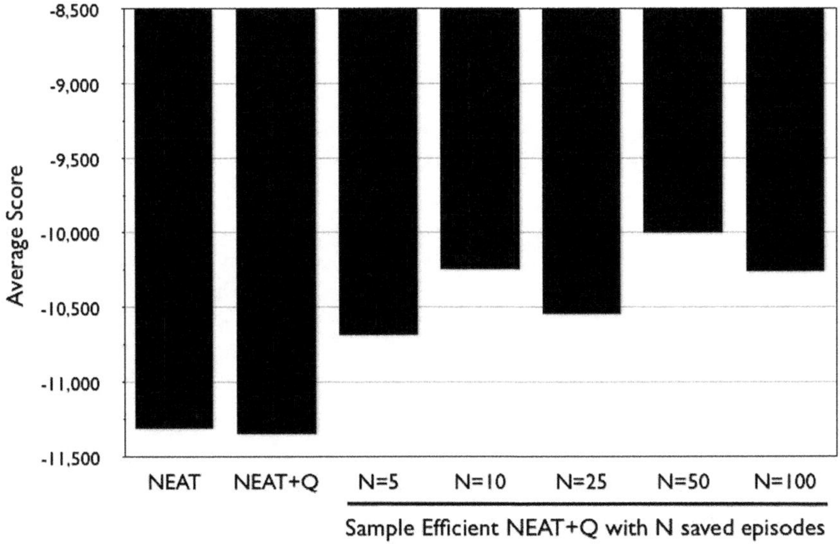

Fig. 5.3 A comparison of the performance of NEAT, NEAT+Q and sample-efficient NEAT+Q with different numbers of saved episodes. Each bar represents the average score after 30,000 episodes and hence is comparable to the right edge of Figure 5.2.

5.3 Discussion

Together, these results clearly demonstrate that sample-efficient evolutionary function approximation can substantially improve performance. Furthermore, it can do so without using more sample episodes than a traditional evolutionary approach. The price for this sample efficiency is increased computational complexity. In practice, this trade-off is likely to be beneficial, since sample experience is typically a much scarcer resource. Even when computational resources are limited, this approach can be useful, as our results demonstrate that even a modest amount of pre-training can significantly improve performance.

Chapter 6
Automatic Feature Selection for Reinforcement Learning

Chapters 3, 4, and 5 introduced methods for automatically optimizing representations for reinforcement learning tasks. However, those methods focus only on the agent's *internal* representation of its solution, i.e., the structure of the mapping from states to actions or from state-action pairs to value estimates. Hence, they still require a human to manually design an *input* representation, i.e., to find a minimal set of features sufficient to describe the agent's current state, a challenge known as the *feature selection* problem. This chapter presents an extension to NEAT designed to automate feature selection in reinforcement learning problems. This extension enables agents to automatically evolve effective representations for their inputs as well as their internal workings.

In many real world tasks, the set of potential inputs that can be fed to the agent is quite large. Feature selection is the process of determining which subset of these inputs should be included to generate the best performance. Doing so correctly can be critical to success. If any important features are excluded, it may be impossible to find an optimal policy. On the other hand, including superfluous inputs can also impede learning. Since each input adds at least one dimension to the search space, even a few extraneous features can be detrimental. However, the consequences of sub-optimal feature selection are not limited just to the learner's performance. If adding inputs costs money (e.g. putting more sensors on a robot), then pruning out unnecessary features can be vital.

Feature selection can often be performed by a human with the appropriate domain expertise. However, in some domains, no one has the requisite knowledge and, even when experts do exist, employing them can be expensive and time consuming. In such domains, automatic feature selection is necessary. Blum and Langley (24) divide feature selection techniques into two categories: *filters* and *wrappers*. Filters (26; 70) analyze the value of a feature set without regard to the learning algorithm that will use those features. Instead, they rely on labeled data. The data is analyzed to determine which features are most useful in distinguishing between the category labels. This approach has been successful but works only in supervised learning tasks. In reinforcement learning scenarios, when no labeled data is available, filtering techniques are not applicable.

S. Whiteson: Adaptive Representations for Reinforcement Learning, SCI 291, pp. 53–64.
springerlink.com

By contrast, wrappers (103; 108) test a feature set by applying it to the given learning algorithm and observing its performance. Labeled examples are not necessary so this approach can be used in reinforcement learning tasks as well. However, it requires a meta-learner to search through the space of feature sets; evaluating any point in that space requires an entire machine learning run of its own. For most real-world problems, this approach is computationally infeasible.

This chapter presents Feature Selective NeuroEvolution of Augmenting Topologies (FS-NEAT) (164), a new learning method that avoids such limitations by incorporating the feature selection problem into the learning task. FS-NEAT searches for good feature sets at the same time as it trains networks that receive those features as input. Hence, it does not depend on human expertise, labeled data sets, or meta-learning.

FS-NEAT is based on NEAT, which evolves both the topology and weights of a neural network. FS-NEAT goes one step further than regular NEAT by learning the network's inputs too. Using evolution, it automatically and simultaneously determines the network's inputs, topology, and weights. Harvey et al. (59) also used neuroevolution to find useful subsets of available features though, unlike FS-NEAT, their system still requires a human to specify the size of that subset in advance.

A critical feature of NEAT is that it begins with networks of minimal topology (i.e., with no hidden nodes and all inputs connected directly to the outputs). As evolution proceeds, NEAT adds links and hidden nodes through mutation. Since only those additions that improve performance are likely to be retained, it tends to find small networks without superfluous structure. Starting minimally also helps NEAT learn more quickly. When networks in its population are small, it is optimizing over a lower-dimensional search space; it jumps to a larger space only when performance in the smaller one stagnates.

FS-NEAT further exploits this same premise. It begins with a population of networks that are even smaller than in regular NEAT. These networks contain no connections at all, not even those connecting inputs to outputs, save those added by an initial mutation step. Hence, they are little more than pools of inputs and outputs. Evolution then proceeds as in regular NEAT, with hidden nodes and links added through mutation. Feature selection occurs implicitly as only those links emerging from useful inputs will tend to survive.

In addition to introducing this novel method, this chapter presents experiments comparing FS-NEAT to regular NEAT in a challenging reinforcement learning domain: an autonomous car racing simulation called RARS (154). The results of these experiments confirm that when some of the available inputs are redundant or irrelevant, FS-NEAT can learn better and faster than regular NEAT. In addition, these results demonstrate that the networks FS-NEAT evolves are smaller and require fewer inputs.

6.1 FS-NEAT

NEAT's initial networks are small but not as small as possible. The structure of the initial networks, in which each input is connected directly to each output, reflects an assumption that all the available inputs are useful and should be connected to the rest of the network. In domains where the input set has been selected by a human expert, this assumption is reasonable. However, in many domains no such expert is available and the input set may contain many redundant or irrelevant features. In such cases, the initial connections used in regular NEAT can significantly harm performance by unnecessarily increasing the size of the search space.

FS-NEAT is an extension of NEAT that attempts to solve this problem by starting even more minimally: with networks having almost no links at all. As in regular NEAT, hidden nodes and links are added through mutation and only those additions that aid performance are likely to survive. Hence, FS-NEAT begins in even lower dimensional spaces than regular NEAT and feature selection occurs implicitly: only those links emerging from useful inputs will tend to survive.

Exactly how should we initialize the population in order to implement this idea? The most minimal initial topology possible would contain no hidden nodes or links at all. However, such networks would not generate any output. Obviously, spending a generation to evaluate a population of such networks would be wasteful. Therefore, for each network in the initial population, FS-NEAT randomly selects an input and an output and adds a link connecting them. Figure 6.1 compares the initial network topologies of regular NEAT and FS-NEAT. After the initial population is generated, FS-NEAT behaves exactly like regular NEAT.

In most tasks, FS-NEAT's initial networks will lack the structure necessary to perform well. However, some will likely connect a relevant input to an output in a useful way and hence outperform their peers. Such early distinctions provide an initial gradient to the evolutionary search. Complexification then drives that search towards networks that use the most appropriate inputs, topology and weights.

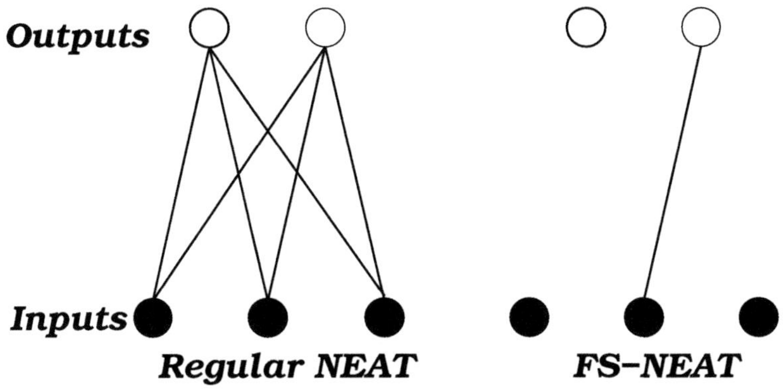

Fig. 6.1 Examples of initial network topologies for both regular NEAT and FS-NEAT. In regular NEAT, networks in the initial population have all inputs connected directly to all outputs. In FS-NEAT, those networks have one link connecting a randomly selected input and output.

Since FS-NEAT incorporates the feature selection problem into the learning task itself, it avoids the need for expensive meta-learners used by wrappers. In addition, since it does not rely on labeled data like filters do, it can be applied to reinforcement learning problems. The next section describes one such application.

6.2 Testbed Domain

The experiments presented in this chapter were conducted in the Robot Auto Racing Simulator (RARS) (154). This domain was selected because of NEAT's previous success evolving controllers for it (136) and because the available inputs (described below) pose a natural feature selection challenge. FS-NEAT could in principle be applied to many other domains, including the mountain car and server job scheduling tasks employed in previous chapters, if irrelevant and redundant features were added. Such experiments are left for future work.

RARS is a Java-based program that uses a two-dimensional model to simulate cars racing around a track. The simulation is quite realistic and takes into account effects such as skidding and traction. In addition, RARS models the noise that occurs in real-world effectors. For example, the coefficient of friction is stochastic such that the effect of trying to accelerate is not entirely predictable. The goal in this domain is to develop a controller that can race an automobile around the track as quickly as possible without damaging it.

The RARS simulator offers a plethora of raw data about the car's immediate environment. This data was consolidated into a rangefinder system, shown in Figure 6.2, that projects rays at different angles relative to the car's current heading. These rays measure the distance from the car to the edge of the road, which allows the agent to estimate its position in the road and perceive upcoming curves. This sensor system creates a very typical feature selection problem. How many rangefinders does the controller need in order to drive the car most effectively? If too few are included, the

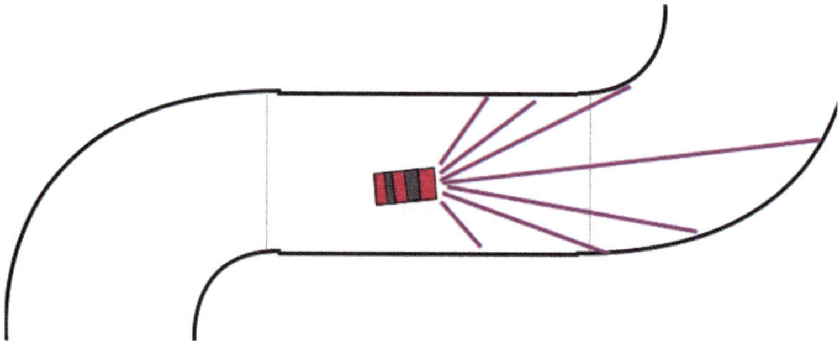

Fig. 6.2 The range finder sensor system in RARS. A set of rays (seven in this case) are projected at different angles to allow the agent to estimate its position in the road and perceive upcoming curves.

networks NEAT evolves will not have enough information to master the task. If too many are included, NEAT will be forced to search in an unnecessarily large search space, which may substantially reduce its performance.

To test the ability of FS-NEAT to automatically address this problem, the networks are provided with a set of 80 rangefinders (evenly distributed across the 180 degree range in front of the car), which we expect to be more than necessary. In addition, another 80 irrelevant inputs are included, each of which supplies random numbers drawn uniformly from the range $[0,1]$. The number of irrelevant and redundant features was selected to ensure a challenging feature selection problem. In Section 6.3, we examine the relative performance of FS-NEAT and NEAT as the number of irrelevant and redundant features varies.

Finally, there is one input specifying the vehicle's current velocity and one bias unit, for a total of 162 inputs. If FS-NEAT can automatically discover a useful subset of these inputs, it should outperform regular NEAT, which is forced to use all 162.

In addition to these inputs, the networks have two outputs: one specifying the agent's desired speed and the other specifying the agent's desired heading. In our experiments, a trial consists of 2000 timesteps on a standard RARS track called "clkwis," shown in Figure 6.3. This track was selected because it is small enough to allow efficient evaluations but still captures a wide range of driving challenges (i.e. straight sections, turns, and an S-shaped curve). During each timestep, input from the environment is fed into the network controlling the car. The network is then activated once and the values of the outputs are used to adjust the vehicle's heading

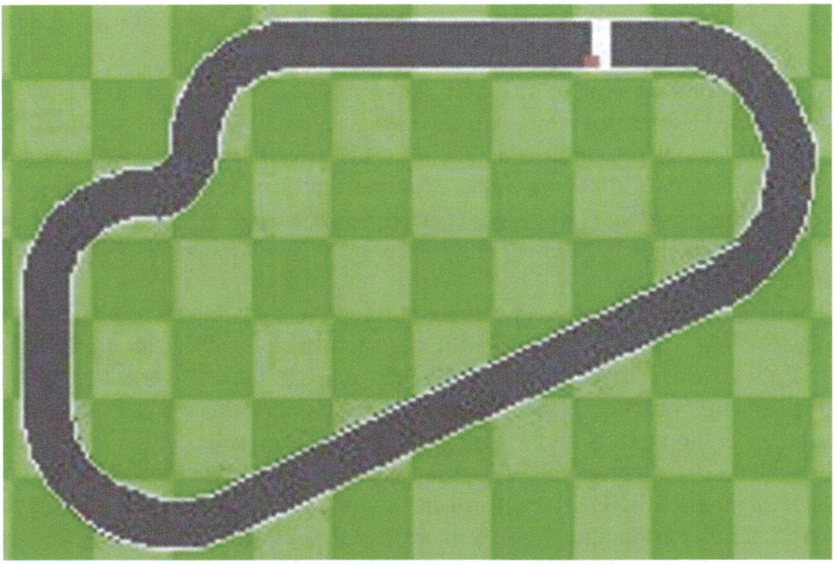

Fig. 6.3 The "clkwis" track used in the FS-NEAT experiments. It captures a wide range of driving challenges (i.e. straight sections, turns, and an S-shaped curve).

and speed. At the end of each trial a score is computed as $S = 2d - b$, where d is the distance traveled and b is a damage penalty computed internally by RARS as a function of the time the vehicle spends off the track. Since the simulation is noisy, each fitness evaluation in NEAT consists of 10 trials; the agent's fitness is the average of the scores received in these trials. Table 6.1 provides more details on the NEAT parameters used in these experiments, which were selected to match those used in previous research about applying NEAT to RARS (136).

Table 6.1 The NEAT parameters used in the experiments described in this chapter. Stanley and Miikkulainen (137) describe the semantics of these parameters in detail.

Parameter	Value	Parameter	Value	Parameter	Value
weight-mut-power	0.5	recur-prop	0.0	disjoint-coeff (c_1)	1.0
excess-coeff (c_2)	1.0	mutdiff-coeff (c_3)	2.0	compat-threshold	3.0
age-significance	1.0	survival-thresh	0.2	mutate-only-prob	0.25
mutate-link-weights-prob	0.9	mutate-add-node-prob (m_n)	0.02	mutate-add-link-prob (m_l)	0.1
interspecies-mate-rate	0.05	mate-multipoint-prob	0.6	mate-multipoint-avg-prob	0.4
mate-singlepoint-prob	0.0	mate-only-prob	0.2	recur-only-prob	0.0
pop-size (p)	100	dropoff-age	1000	newlink-tries	50
babies-stolen	0	num-compat-mod	0.3	num-species-target	6

6.3 Results

Using this setup, we performed experiments comparing regular NEAT to FS-NEAT. For each method, we conducted 10 runs, each of which ran for 200 generations. The results are summarized in Figure 6.4. Each line in the graph represents the score received by the best network from each generation, averaged over all 10 runs. The graph demonstrates that when some of the available inputs are redundant or irrelevant, FS-NEAT can learn better networks and learn them faster than regular NEAT. In this graph and all those presented below, a Student's t-test verified, with 95% confidence, the statistical significance of the difference between FS-NEAT and regular NEAT.

Figure 6.5 shows, for the same experiments, how many inputs have at least one connection emerging from them in the best network of each generation. Regular NEAT always uses all 162 inputs but FS-NEAT finds better networks that use only a small fraction of them. In fact, when FS-NEAT's performance begins to plateau around generation 65, its performance is already 17.5% better than regular NEAT ever achieves, at which point its best network has on average only 10% as many connected inputs. FS-NEAT's performance continues to creep up slowly after generation 65, improving another 4.6% by generation 200, at which point its best network has on average 22.9% as many inputs as regular NEAT.

Figure 6.6 shows, for the same experiments, the size of the best network from each generation, where size is simply the total number of nodes (only connected

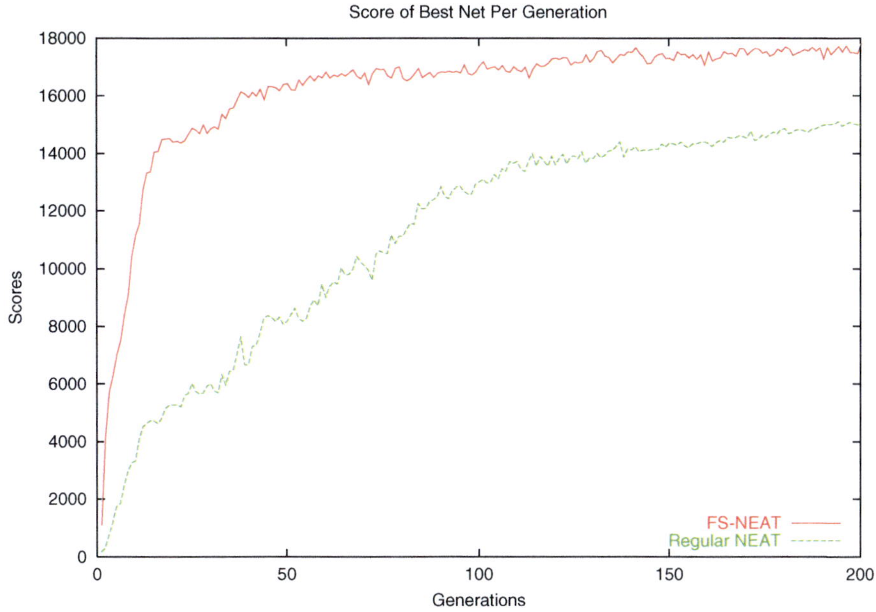

Fig. 6.4 A comparison of the performance of regular NEAT and FS-NEAT in the RARS domain with 162 available inputs, 80 of which are irrelevant to the task. Each line in the graph represents the score received by the best network from each generation, averaged over all 10 runs. By learning appropriate feature sets, FS-NEAT learns significantly better networks and learns them faster than regular NEAT.

inputs are counted) plus the total number of links. This graph demonstrates that FS-NEAT evolves substantially smaller networks than regular NEAT does. When FS-NEAT's performance begins to plateau around generation 65, its best network is on average only 9.7% as large as regular NEAT's. When the runs complete at generation 200, FS-NEAT's best network is on average only 18.5% as large as regular NEAT's.

In these experiments, FS-NEAT found high performing networks that use only 16 inputs, which implies that the feature set we supplied to the learners, with 80 rangefinders, was much larger than needed. How would the performance of FS-NEAT relative to regular NEAT change if the initial feature set were closer to ideal? How many redundant and irrelevant features must be present before FS-NEAT provides a significant advantage? Does FS-NEAT's performance improvement continue to increase as the feature set gets larger? To address these questions, we conducted several additional experiments with feature sets of different sizes. These experiments use the setup described above but instead of 80 rangefinders they include 5, 20, 40, or 160 rangefinders. In each case, the rangefinders are matched with an equal number of irrelevant inputs. Adding the velocity and bias inputs yields initial feature sets of size 12, 42, 82, and 322. For each size and for each method, we conducted 10 runs, each of which ran for 200 generations.

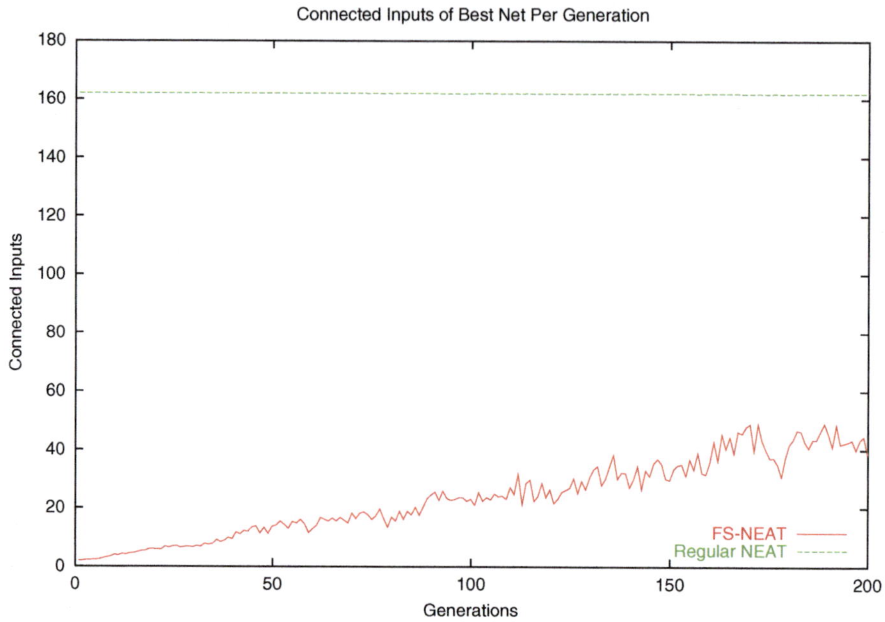

Fig. 6.5 A comparison of the number of inputs used by regular NEAT and FS-NEAT in the RARS domain with 162 available inputs. Each line in the graph represents the number of inputs with at least one connection emerging from them in the best network of each generation. Regular NEAT always uses all 162 inputs but FS-NEAT evolves better networks that uses significantly fewer of them.

Figure 6.7 summarizes the results of these experiments by showing, for each method and feature set size, the performance of the best network in the entire run, averaged over all ten runs. Even when the initial feature set contains only 12 inputs, FS-NEAT still performs better. As the size of the feature set grows, the performance of regular NEAT deteriorates. By contrast, the performance of FS-NEAT remains nearly constant even as the feature selection task it faces becomes ever more difficult.

Figure 6.8 compares the number of connected inputs in the best network in the entire run, averaged over all ten runs. Regular NEAT always uses all available inputs while FS-NEAT learns to use much smaller subsets. Even as the size of the feature set grows, the number of inputs used by FS-NEAT's best networks stays nearly constant. Similarly, Figure 6.9 compares the sizes of these same networks. The size of regular NEAT's best networks increases linearly with respect to the number of available features, whereas FS-NEAT's best networks stay nearly constant in size. Therefore, FS-NEAT consistently uses features sets with many fewer extraneous inputs than regular NEAT and, in so doing, finds better solutions faster.

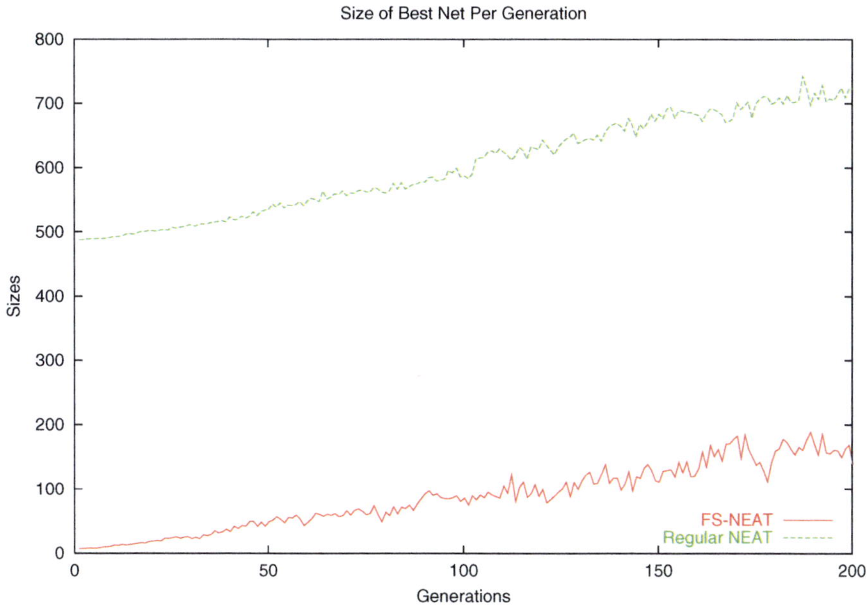

Fig. 6.6 A comparison of the size of the networks evolved by regular NEAT and FS-NEAT in the RARS domain with 162 available inputs. Each line in the graph represents the number of nodes (only connected inputs are counted) plus the number of links in the best network of each generation. FS-NEAT evolves significantly smaller networks than regular NEAT does.

6.4 Discussion

The empirical results presented in this chapter demonstrate that when some of the available inputs are redundant or irrelevant, FS-NEAT can learn better networks and learn them faster than regular NEAT. In addition, the networks it learns are smaller and use fewer inputs. These results are consistent across feature sets of different sizes.

One interesting question raised by these results is why the size and number of inputs used by FS-NEAT do not plateau. For example, Figure 6.4 shows that performance improvements mostly level off by generation 65. However, Figures 6.5 and 6.6 show that the size and number of inputs used by FS-NEAT's best networks continue to grow linearly through generation 200. Should we not expect them to plateau also once the "right" size has been found? Counterintuitively, the answer is no. The goal of both NEAT and FS-NEAT is to determine the right complexity to solve a given task. Hence, when performance at a certain complexity plateaus, these algorithms proceed to explore at higher complexities. In these experiments, that exploration pays few dividends after generation 65.

Nonetheless, even given such exploration, we would still expect to see size plateau if there were a strong selective pressure against larger networks since none

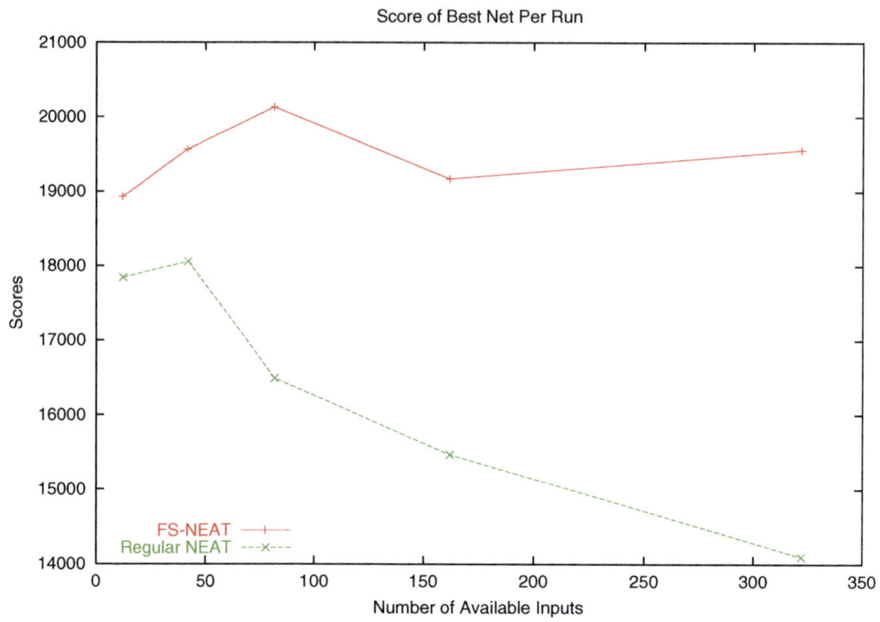

Fig. 6.7 A comparison of the performance of regular NEAT and FS-NEAT across feature sets of different sizes. Each line in the graph represents the score received by the best network in the entire run, averaged over all 10 runs. The performance of regular NEAT gets significantly worse as the feature set gets larger whereas the performance of FS-NEAT stays nearly constant.

of these networks would likely become generation champions. The fact that they do implies that FS-NEAT is not completely intolerant of redundant and irrelevant inputs. This behavior makes sense because the presence of such inputs may not be harmful if, for example, NEAT can learn to set the weights emerging from them close to zero. In this respect, FS-NEAT behaves exactly as we would wish: it selects against large networks only when their size presents a significant disadvantage to the learner.

In evolutionary search, it is critical that the fitness of the initial population have some variance: unless some individuals are more promising than others, progress is unlikely. This issue is of particular concern in FS-NEAT since its initial population consists of degenerate networks that are almost completely disconnected. While the experiments presented in this chapter verify that FS-NEAT consistently finds an initial gradient for learning, those experiments tested only one population size: 100. We wondered if the relative performance of FS-NEAT would deteriorate for smaller populations since the probability of finding an initial promising network would decrease. However, this problem does not occur in the RARS domain. In fact, informal experiments with different population sizes indicate that both regular NEAT and FS-NEAT perform robustly with populations as small as 25 and that FS-NEAT retains

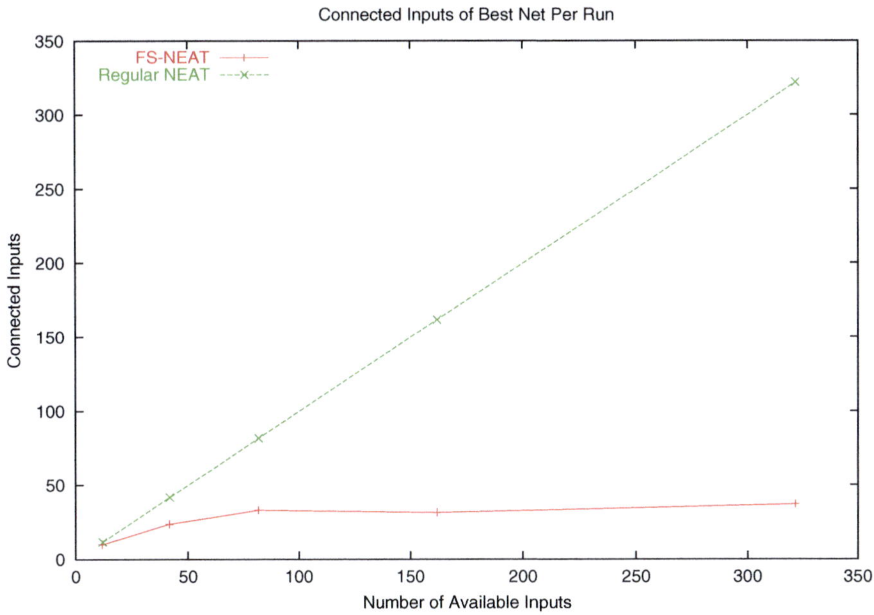

Fig. 6.8 A comparison of the number of connected inputs in regular NEAT and FS-NEAT across feature sets of different sizes. Each line in the graph represents the number of inputs with at least one connection in the best network of the entire run, averaged over all 10 runs. Regular NEAT always uses all available inputs while FS-NEAT learns to use significantly smaller subsets.

its substantial advantage over regular NEAT. Hence, at least in RARS, FS-NEAT's smaller initial networks seem *more* likely to point evolution in the right direction.

In other settings, however, the lack of initial gradient may be a serious problem. For example, when FS-NEAT is combined with NEAT+Q, to evolve both input and internal representations of neural network function approximators, performance is poor, perhaps for this reason (see Section 9.2.1 for a discussion of this negative result). Nonetheless, FS-NEAT has also been successfully applied in a domain very different from RARS, namely distributed instruction placement in compilers (37). Hence, the initial gradient required for FS-NEAT to succeed is not unique to the RARS domain.

The most revealing test of FS-NEAT's robustness is how its performance changes when the size of the initial feature set increases. As this set gets larger, feature selection becomes more important, as confirmed by the decline of regular NEAT's performance in Figure 6.7. FS-NEAT's performance, by contrast, does not decline at all. Most strikingly, the size and number of inputs used by FS-NEAT's best networks remains approximately constant across different feature set sizes, whereas regular NEAT's networks grow ever larger. Together, these results suggest that the efficacy of FS-NEAT scales well to large feature selection problems.

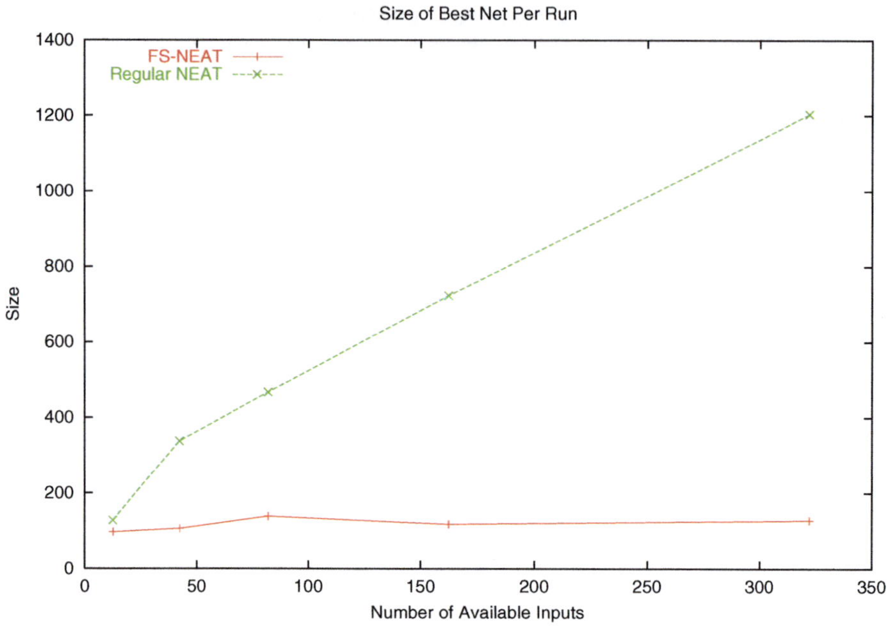

Fig. 6.9 A comparison of network size in regular NEAT and FS-NEAT across feature sets of different sizes. Each line in the graph represents the size of the best network of the entire run, averaged over all 10 runs. Regular NEAT's best networks increase in size significantly as the number of available features grows, whereas FS-NEAT's best networks stay nearly constant in size.

Chapter 7
Adaptive Tile Coding

Chapters 4 and 5 demonstrate how an agent can automatically adapt the representation of neural network function approximators. This chapter extends that work by introducing adaptive methods for a different type of function approximator, namely tile coding. Extending adaptive methods beyond neural networks is important because, while neural networks are a powerful representation, they are not a panacea. On the contrary, they have some significant drawbacks. Perhaps chief among these is their inscrutability. Even when neural networks perform well, their inner workings are typically difficult or impossible for a human to understand.

This "black box" quality means that the cause of poor performance is often difficult to diagnose. Hence, it is not feasible for the agent to *reason* about the inadaquecies of its representation and how best to remedy them. Instead, it can only *search* for a good representation, which is why Chapter 4 focuses on an evolutionary approach to finding good representations. While Chapter 5 demonstrates that such an approach can be made sample-efficient, testing each candidate representation remains expensive.

By contrast, the behavior of linear representations such as tile codings are typically much easier to interpret. Unlike neural networks, the effects of any particular weight are contained in a particular region of the state space. Furthermore, changes to the representation (e.g., splitting tiles in two) have consequences that are largely predictable. Hence, an agent, by analyzing its own behavior, can reason about how to improve its representation without requiring expensive search. This chapter presents *adaptive tile coding*, a novel method for doing so.

Tile coding, which forms a piecewise-constant approximation of the value function, requires a human designer to choose the size of each tile in each dimension of the state space. Adaptive tile coding automates this process by starting with large tiles and making them smaller during learning by splitting existing tiles in two. Beginning with simple representations and refining them over time is a strategy that has proven effective for NEAT and NEAT+Q, as well as other function approximators (36; 102). In addition to automatically finding good representations, this approach gradually reduces the function approximator's level of generalization over time, a factor known to critically affect performance in tile coding (126).

S. Whiteson: Adaptive Representations for Reinforcement Learning, SCI 291, pp. 65–76.
springerlink.com

To succeed, adaptive tile coding must make smart decisions about which tiles to split and along which dimension. This chapter compares two different criteria for prioritizing potential splits. The *value criterion* estimates how much the value function will change if a particular split occurs. By contrast, the *policy criterion* estimates how much the policy will change if a given split occurs.

Empirical results in two benchmark reinforcement learning tasks demonstrate that the policy criterion is more effective than the value criterion. The results also verify that adaptive tile coding can automatically discover representations that yield approximately optimal policies and that the adaptive approach's speed of learning is competitive with the best fixed tile-coding representations.

7.1 Background

This section briefly describes tile coding representations and how they are used to approximate value functions. For simplicity, this chapter focuses on MDPs that are continuous but deterministic, though in principle the methods presented could be extended to stochastic domains. Hence, the transition dynamics are described by $T : S \times A \rightarrow S$ such that an agent in state $s \in S$ that takes action $a \in A$ will transition to state $T(s,a)$. As in previous work on adaptive function approximation (36; 55; 102), we also assume the agent has a model of its environment (i.e., T and R are known). Hence, the agent need only learn V^*, not Q^*.

7.1.1 Tile Coding

In tile coding (4), a piecewise-constant approximation of the optimal value function is represented by a set of exhaustive partitions of the state space called *tilings*. Typically, the tilings are all partitioned in the same way but are slightly offset from each other. Each element of a tiling, called a *tile*, is a binary feature activated if and only if the given state falls in the region delineated by that tile. Figure 7.1 illustrates a tile-coding scheme with two tilings.

The value function that the tile coding represents is determined by a set of weights, one for each tile, such that

$$V(s) = \sum_{i=1}^{n} b_i(s) w_i$$

where n is the total number of tiles, $b_i(s)$ is the value (0 or 1) of the ith tile given state s, and w_i is the weight of that tile. In practice, it is not necessary to sum over all n tiles since only one tile in each tiling is activated for a given state. Given m tilings, we can simply compute the indices of the m active tiles and sum their associated weights.

Given a model of the MDP as described above, we can update the value estimate of a given state s by computing $\Delta V(s)$ using dynamic programming:

$$\Delta V(s) = \max_a [R(s,a) + \gamma V(T(s,a))] - V(s)$$

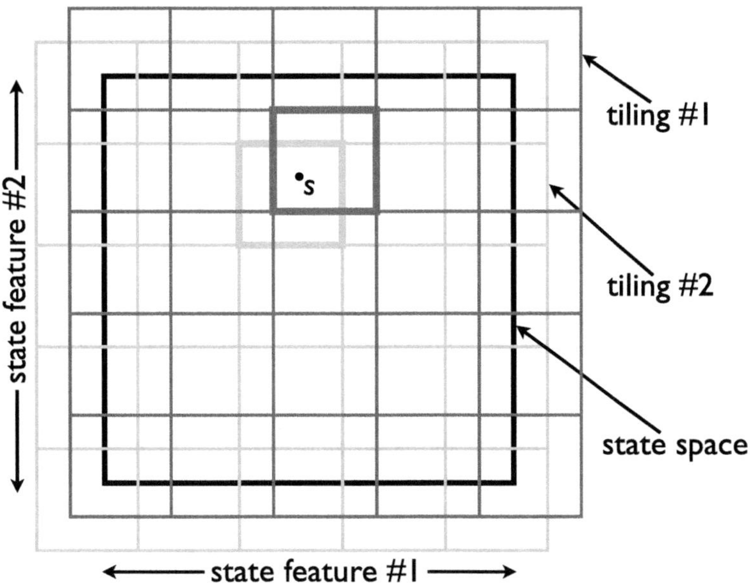

Fig. 7.1 An example of tile coding with two tilings. Thicker lines indicate which tiles are activated for the given state s.

and adjusting each weight so as to reduce $\Delta V(s)$:

$$w_i \leftarrow w_i + \frac{\alpha}{m} b_i(s) \Delta V(s)$$

where α is a learning rate parameter. As before, it is not necessary to update all n weights, only the m weights associated with tiles activated by state s. Algorithm 8 shows a simple way to learn an approximation of the optimal value function using

---

Algorithm 8 TILE-CODING($S, A, T, R, \alpha, \gamma, m, n$)

---

for $i \leftarrow 1$ to m **do**
    Initialize tiling i with n/m tiles
    for $j \leftarrow 1$ to n/m **do**
        Initialize tile j with zero weight
repeat
    $s \leftarrow$ random state from S
    $\Delta V(s) \leftarrow \max_a [R(s,a) + \gamma V(T(s,a))] - V(s)$
    for $i \leftarrow 1$ to m **do**
        $w \leftarrow$ weight of ACTIVE-TILE(s)
        $w \leftarrow w + \frac{\alpha}{m} \Delta V(s)$
until time expires

---

tile coding. The function ACTIVE-TILE returns the tile in the given tiling activated by the given state. If only one tiling is used, then there is a trade-off between speed and precision of learning. Smaller tiles yield more precise value estimates but take longer to learn since those estimates generalize less broadly. Multiple tilings can avoid this trade-off, since more tilings improve resolution without reducing capacity for generalization.

7.2 Method

Tile coding is a simple, computationally efficient method for approximating value functions that has proven effective (144; 140). However, it has two important limitations.

The first limitation is that it requires a human designer to correctly select the width of each tile in each dimension. While in principle tiles can be of any size and shape, they are typically axis-aligned rectangles whose widths are uniform within a given dimension. Selecting these widths appropriately can mean the difference between fast, effective learning and catastrophically poor performance. If the tiles are too large, value updates will generalize across regions in S with disparate values, resulting in poor approximations. If the tiles are too small, value updates will generalize very little and learning may be infeasibly slow.

The second limitation is that the degree of generalization is fixed throughout the learning process. The use of multiple tilings makes it possible to increase resolution without compromising generalization, but the degree of generalization never changes. This limitation is important because recent research demonstrates that the best performance is possible only if generalization is gradually reduced over time (126). Intuitively, broad generalization at the beginning allows the agent to rapidly learn a rough approximation; less generalization at the end allows the agent to learn a more nuanced approximation.

This section presents *adaptive tile coding*, a novel function approximation method that addresses both of these limitations. The method begins with simple representations and refines them over time, a strategy that has proven effective for NEAT and NEAT+Q, as well as well as piecewise-linear representations based on kd-trees (102), and uniform grid discretizations (36). Adaptive tile coding begins with a few large tiles, and gradually adds tiles during learning by splitting existing tiles. While there are infinitely many ways to split a given tile, for the sake of computational feasibility, our method considers only splits that divide tiles in half evenly. Figure 7.2 depicts this process for a domain with two state features.

By analyzing the current value function and policy, the agent can make smart choices about when and where to split tiles, as detailed below. In so doing, it can automatically discover an effective representation that devotes more resolution to critical regions of S, without the aid of a human designer. Furthermore, learning with a coarse representation first provides a natural and automatic way to reduce

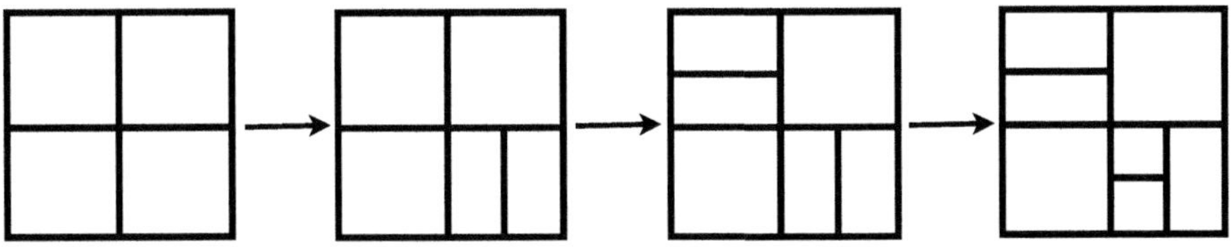

Fig. 7.2 An example of how tiles might be split over time using adaptive tile coding

generalization over time. As a result, multiple tilings are no longer necessary: a single, adaptive tiling can provide the broad generalization needed early in learning and the high resolution needed later on. The remainder of this section addresses two critical issues: when and where to split tiles.

7.2.1 When to Split

Correctly deciding when to split a tile can be critical to performance. Splitting a tile too soon will slow learning since generalization will be prematurely reduced. Splitting a tile too late will also slow learning, as updates will be wasted on a representation with insufficient resolution to further improve value estimates. Intuitively, the agent should learn as much as possible with a given representation before refining it. Hence, it needs a way to determine when learning has plateaued.

One way to do so is by tracking *Bellman error* (i.e., ΔV). As long as V is improving, $|\Delta V|$ will tend to decrease over time. However, this quantity is extremely noisy, since updates to different tiles may differ greatly in magnitude and updates to different states within a single tile can move the value estimates in different directions. Hence, a good rule for deciding when to split should consider Bellman error but be robust to its short-term fluctuations.

We use the following heuristic. For each tile, the agent tracks the lowest $|\Delta V|$ occurring in updates to that tile. It also maintains a global counter u, the number of updates occurring since the updated tile had a new lowest $|\Delta V|$ (each update either increments u or resets it to 0). When u exceeds a threshold parameter p, the agent decides that learning has plateaued and selects a tile to split. In other words, a split occurs after p consecutive updates fail to produce a new tile-specific lowest $|\Delta V|$.[1] Hence, the agent makes a *global* decision about when learning has finished, since $|\Delta V|$ may temporarily plateau in a given tile simply because the effects of updates to other tiles have not yet propagated back to it.

[1] There are many other ways to determine when learning has plateaued. For example, in informal experiments, we applied linear regression to a window of recent $|\Delta V|$ values. Learning was deemed plateaued when the slope of the resulting line dropped below a small threshold. However, this approach proved inferior in practice to the one described above, primarily because performance was highly sensitive to the size of the window.

7.2.2 Where to Split

Once the agent decides that learning has plateaued, it must decide which tile to split and along which dimension.[2] This section presents two different approaches, one based on expected changes to the value function and the other on expected changes to the policy. Both require the agent to maintain *sub-tiles*, which estimate, for each potential split, what weights the resulting tiles would have. Since each state is described by k state features, each tile has $2k$ sub-tiles.

When a new tile is created, its sub-tile weights are initialized to zero. When the agent updates state s, it also updates the k sub-tiles that are activated by s, using the same rule as for regular weights, except that the update is computed by subtracting the relevant sub-tile weight (rather than the old value estimate) from the target value:

$$\Delta w_d(s) = \max_a[R(s,a) + \gamma V(T(s,a))] - w_d(s)$$

where $w_d(s)$ is the weight of the sub-tile resulting from a split along dimension d activated by state s. Algorithm 9 describes the resulting method, with regular weight updates in lines 8–9 and sub-tile weight updates in lines 10–13. In line 19, the agent selects a split according to one of the criteria detailed in the remainder of this section.

Algorithm 9 ADAPTIVE-TILE-CODING($S, A, T, R, k, \alpha, \gamma, n, p$)

1: $u \leftarrow 0$
2: Initialize one tiling with n tiles
3: **for** $i \leftarrow 1$ to n **do**
4:     Initialize ith tile and $2k$ sub-tile weights to zero
5: **repeat**
6:     $s \leftarrow$ random state from S
7:     $\Delta V(s) \leftarrow \max_a[R(s,a) + \gamma V(T(s,a))] - V(s)$
8:     $w \leftarrow$ weight of tile activated by s
9:     $w \leftarrow w + \alpha \Delta V(s)$
10:     **for** $d \leftarrow 1$ to k **do**
11:         $w_d \leftarrow$ weight of sub-tile w.r.t split along d activated by s
12:         $\Delta w_d = \max_a[R(s,a) + \gamma V(T(s,a))] - w_d$
13:         $w_d \leftarrow w_d + \alpha \Delta w_d$
14:     **if** $|\Delta V| <$ lowest Bellman error on tile activated by s **then**
15:         $u \leftarrow 0$
16:     **else**
17:         $u \leftarrow u + 1$
18:     **if** $u > p$ **then**
19:         Perform split that maximizes value or policy criterion
20:         $u \leftarrow 0$
21: **until** time expires

[2] The agent splits only one tile at a time. It could split multiple tiles but doing so would be similar to simply reducing p.

7.2.2.1 Value Criterion

Sub-tile weights estimate what values the tiles resulting from a potential split would have. Thus, the difference in sub-tile weights indicates how drastically V will change as a result of a given split. Consequently, the agent can maximally improve V by performing the split that maximizes, over all tiles, the value of $|w_{d,u} - w_{d,l}|$, where $w_{d,u}$ and $w_{d,l}$ are, respectively, the weights of the upper and lower sub-tiles of a potential split d. Using this *value criterion* for selecting splits will cause the agent to devote more resolution to regions of S where V changes rapidly (where generalization will fail) and less resolution to regions where it is relatively constant (where generalization is helpful).

7.2.2.2 Policy Criterion

The value criterion will split tiles so as to minimize error in V. However, doing so will not necessarily yield maximal improvement in π. For example, there may be regions of S where V^* changes significantly but π^* is constant. Hence, the most desirable splits are those that enable the agent to improve π, regardless of the effect on V. To this end, the agent can estimate, for each potential split, how much π would change if that split occurred.

When updating a state s, the agent iterates over the $|A|$ possible successor states to compute a new target value. For each dimension d along which each successor state s' could be split, the agent estimates whether $\pi(s)$ would change if the tile activated by s' were split along d, by computing the expected change in $V(s')$ that split would cause:

$$\Delta V_d(s') = w_d(s') - V(s')$$

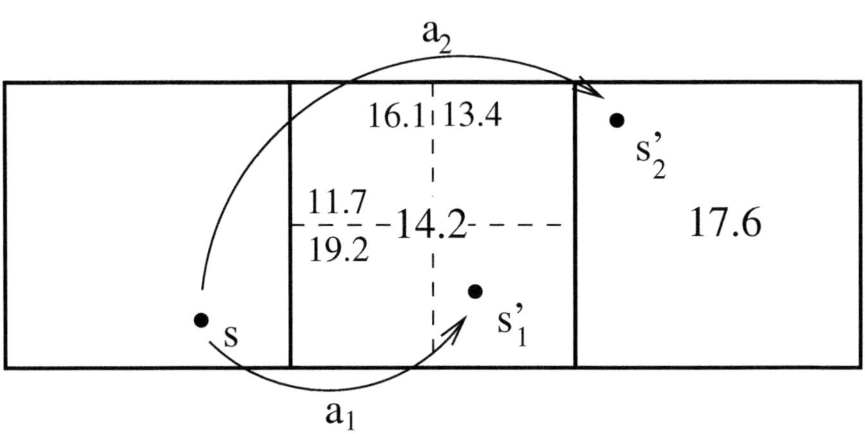

Fig. 7.3 An agent updates state s, from which each action a_i leads to successor state s'_i. The figure shows the tiles, including weights, that these successor states fall in and shows sub-tile weights for the middle tile. Though $\pi(s) = 2$, a horizontal split to the middle tile would make $\pi(s) = 1$ (since $19.2 > 17.6$), incrementing c_d for that split.

If changing $V(s')$ by $\Delta V_d(s')$ would alter $\pi(s)$, then the agent increments a counter c_d, which tracks *changeable actions* for potential split d in the tile activated by s' (see Figure 7.3). Hence, the agent can maximize improvement to π by performing the split that maximizes the value of c_d over all tiles. Using this *policy criterion*, the agent will focus splits on regions where more resolution will yield a refined policy.

7.3 Testbed Domains

In addition to the mountain car domain described in Section 3.4.1, we evaluate adaptive tile coding in puddle world (144), another benchmark reinforcement learning domains whose continuous state features necessitate function approximation. In puddle world, a simulated robot is placed in a random location within a two-dimensional unit square, depicted in Figure 7.4. The robot must navigate this space to reach a goal region which lies in the upper right corner of the square. To do so efficiently, the robot must avoid two puddle regions, which it incurs negative reward for passing through.

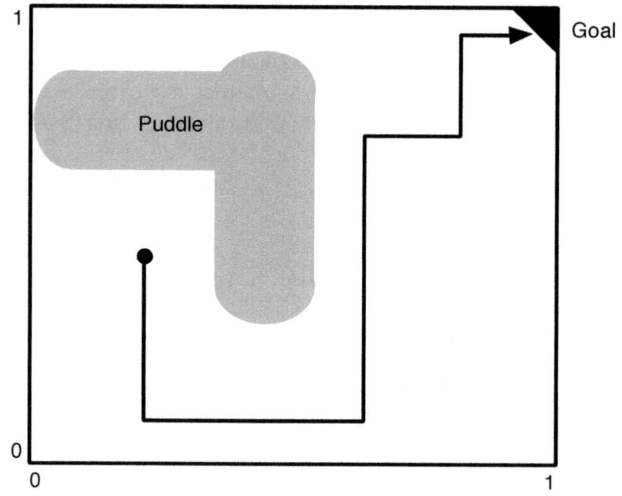

Fig. 7.4 The puddle world domain, in which a robot must navigate a two-dimensional space to reach a goal while avoiding two overlapping puddles

The agent's state is described by two continuous state variables x and y, corresponding to its position in the square. The agent has four actions available to it, each of which moves the robot up, down, left, or right by 0.05, though the robot cannot travel outside the square. Noise drawn from a Gaussian distribution with a mean of 0.0 and standard deviation of 0.01 is added to the distance covered by each action.

The goal region consists of the set of states for which $x+y>1.9$. Since we want the robot to reach the goal as quickly as possibly, the agent incurs a reward of -1 for each time step. In addition, since we want the robot to avoid the puddles, an additional negative reward occurs when the robot is in a puddle. The reward is -400 times the distance inside the puddle.

7.4 Results

To evaluate adaptive tile coding, we tested its performance in the mountain car and puddle world domains. The value and policy criteria were tested separately, with 25 independent trials for each method in each domain. In each trial, the method was evaluated during learning by using its current policy to control the agent in test episodes. The agent took one action for each update that occurred (i.e., one iteration of the repeat loop in Algorithms 8 and 9). Note that since the agent learns from a model, these test episodes do not affect learning; their sole purpose is to evaluate performance. The following parameter settings were used in all trials: $\alpha=0.1$, $\gamma=0.999$, $n=4$ (2x2 initial tilings), and $p=50$.

Next, we tested 18 different fixed tile-coding representations, selected by choosing three plausible values for the number of tilings $m\in\{1,5,10\}$ and six plausible values for the number of tiles n such that the tiles per feature $\sqrt[k]{n/m}\in\{5,10,25,50,100,250\}$, where $k=2$ is the number of state features in each domain. We tested each combination of these two parameters with $\alpha=0.1$ and $\gamma=0.999$ as before. We conducted 5 trials at each of the 18 parameter settings and found that only six in mountain car and seven in puddle world were able to learn good policies (i.e., average reward per episode >-100) in the time allotted.

Finally, we selected the three best performing fixed settings and conducted an additional 25 trials. Figure 7.5 shows the results of these experiments by plotting, for each domain, the uniform moving average reward accrued over the last 500

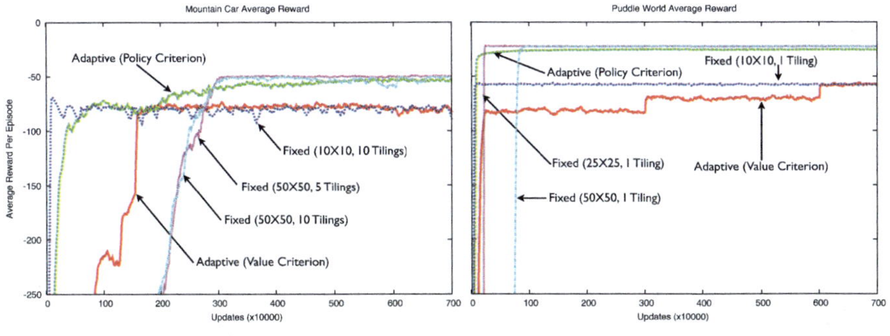

Fig. 7.5 Average reward per episode in both mountain car and puddle world of the adaptive approach with value or policy criterion, compared to the best-performing fixed representations.

episodes for each adaptive approach and the best fixed approaches, averaged over all 25 trials for each method.

7.5 Discussion

The variation in performance among the best fixed representations demonstrates that the choice of representation is a crucial factor in both the speed and quality of learning. Without *a priori* knowledge about what representations are effective in each task, both versions of the adaptive method consistently learn good policies, while only a minority of the fixed representations do so. Furthermore, when the policy criterion was used, the adaptive method learned approximately optimal policies in both domains, at speeds that are competitive with the best fixed representations.

While there are fixed representations that learn good policies as fast or faster than the adaptive approach (10x10 with 10 tilings in mountain car and 10x10 with 1 tiling in puddle world), those representations do not go on to learn approximately optimal policies as the adaptive approach does. Similarly, there are fixed representations that learn approximately optimal policies faster than the adaptive approach (50x50 with 10 tilings in mountain car and 25x25 with 1 tiling in puddle world), but those representations take significantly longer to learn good policies.

Furthermore, the fixed representations that learn good policies fastest are not the same as those that learn approximately optimal policies and are different in the two domains. By contrast, the adaptive method, with a single parameter setting, rapidly learns approximately optimal policies in both domains. Overall, these results confirm the efficacy of the adaptive method and suggest it is a promising approach for improving function approximation when good representations are not known *a priori*.

To better understand why the adaptive method works, we took the best representations learned with the policy criterion, reset all the weights to zero, and restarted learning with splitting turned off. The restarted agents learned much more slowly than the adaptive agents that began with coarse representations and bootstrapped their way to good solutions. This result suggests that the adaptive approach learns well, not just because it finds good representations, but also because it gradually reduces generalization, confirming the conclusions of Sherstov and Stone (126).

The results also demonstrate that the policy criterion ultimately learns better policies than the value criterion. To understand why, we examined the structure of the final representations learned with each approach, as depicted in Figure 7.6. In both domains the value criterion devotes more resolution to regions where V changes most rapidly, as can be seen by comparing the top row of Figure 7.6 with Figure 7.7, which shows typical final value functions learned with the adaptive approach. In mountain car, this region spirals outward from the center, as the agent oscillates back and forth to build momentum. In puddle world, this region covers the puddles, where reward penalties give V a sharp slope, and the area adjacent to the goal. However, those regions do not require fine resolution to represent approximately optimal policies. On the contrary, Figure 7.8, which shows typical final policies learned with the adaptive approach, reveals that π is relatively uniform in those regions.

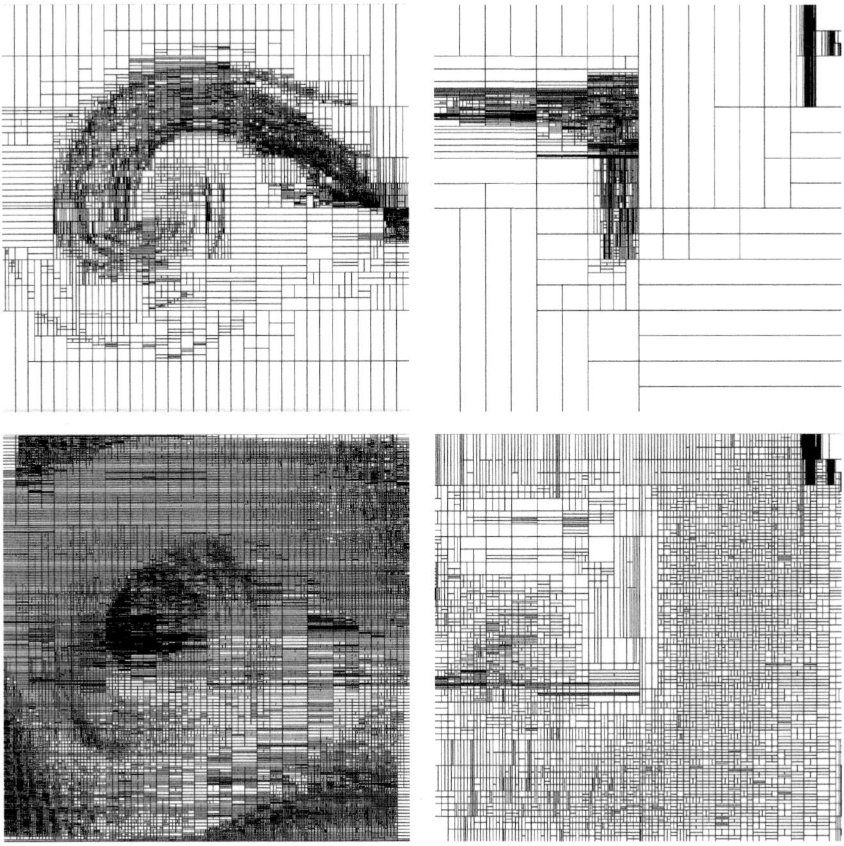

Fig. 7.6 Examples of final tile coding representations learned by the adaptive methods: mountain car in the left column, puddle world in the right column value criterion in the top row, and policy criterion in the bottom row.

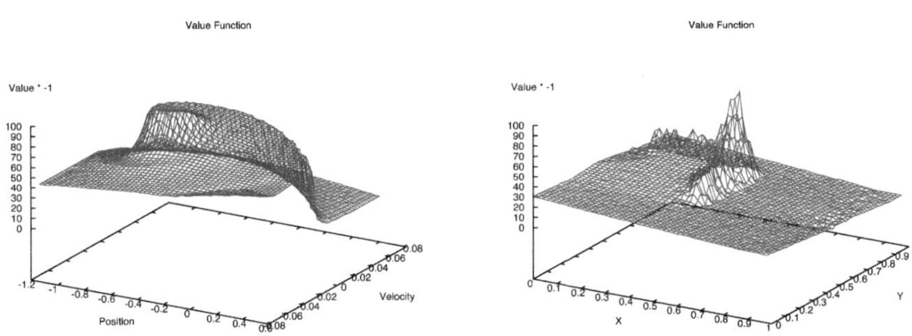

Fig. 7.7 Examples of final value functions learned by adaptive tile coding in both the mountain car (left) and puddle world (right) domains. For greater clarity, the z-axis shows the additive inverse of the value function, i.e., $-V(s)$.

By contrast, the policy criterion devotes more resolution to regions where the policy is not uniform, as can be seen by comparing the bottom row of Figure 7.6 with Figure 7.8. In mountain car, the smallest tiles occur in the center and near each corner, where π is less consistent. In puddle world, the least resolution is devoted to the puddle, where the policy is mostly uniform, and more resolution to the right side, where the "up" and "right" actions are intermingled. Hence, by striving to refine the agent's policy instead of just its value function, the policy criterion makes smarter choices about which tiles to split and consequently learns better policies.

Overall, these results demonstrate that finding the right representation is critical to the success of tile coding function approximators. They also demonstrate that adaptive tile coding can automate this design process. Furthermore, the success of this adaptive approach shows that, for representations like tile codings that are more interpretable than neural networks, adaptive methods can excel without expensive search. See Section 9.3.3 for a comparison of this approach to the search-based methods described in earlier chapters.

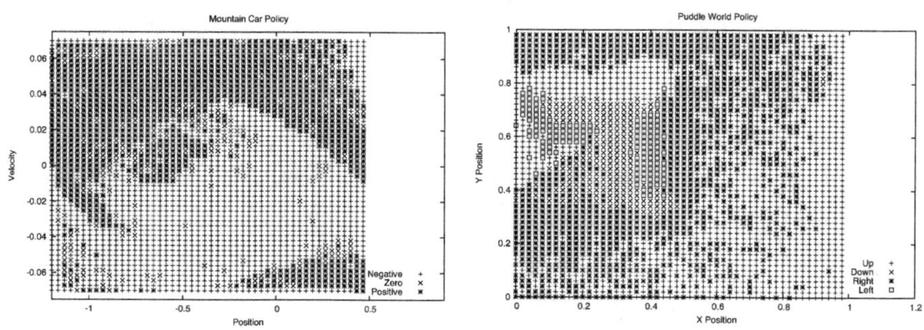

Fig. 7.8 Examples of final policies learned by adaptive tile coding in both the mountain car (left) and puddle world (right) domains

Chapter 8
Related Work

A broad range of previous research is related in terms of both methods and goals to the techniques presented in this book. This chapter presents an overview of that research and discusses the similarities and differences to this work.

Section 8.1 discusses methods for optimizing representations, which are related to evolutionary function approximation (Chapters 4 and 5) and adaptive tile coding (Chapter 7). Section 8.2 presents various approaches to combining evolution (or other policy search methods) with learning, which is also related to evolutionary function approximation. Section 8.3 reviews work addressing the trade-off between exploration and exploitation, which is related to on-line evolutionary computation (Chapter 3), and Section 8.4 reviews work on feature selection, which is related to FS-NEAT (Chapter 6).

8.1 Optimizing Representations

This section reviews previous work on the problem of finding effective representations, which has been studied extensively in the contexts of supervised learning, reinforcement learning, and evolutionary computation. It also discusses how these methods relate to the representation-learning methods introduced in this book (in Chapters 4, 5, and 7).

8.1.1 Supervised Learning

Unlike reinforcement learning, *supervised learning* (96). aims to approximate a function given example input-output pairs. Such *labeled* training data can be statistically analyzed to deduce which representations might best approximate the function.

Perhaps the most well-known methods that employ this approach are the ID3 (116) and C4.5 (117) algorithms for learning *decision trees*. A decision tree represents a discrete-valued function such that each node corresponds to an input feature and each branch emerging from that node corresponds to a value for that feature.

S. Whiteson: Adaptive Representations for Reinforcement Learning, SCI 291, pp. 77–94.
springerlink.com

The leaves of the tree are labeled with the output of the function given the input represented by the path from the root to that leaf. The ID3 and C.45 algorithms perform a top-down greedy search for a decision tree whose structure is appropriate for the given training data. At each step, the algorithm determines which feature to test at the current node of the tree. It selects the feature that maximizes the *information gain*, or decrease in *entropy* in the training set. Hence, statistical analysis of the training set makes it possible to automatically find effective decision tree representations for supervised learning, though the reliance on greedy search means it converges only to a local optimum.

Methods that optimize representations for supervised learning also exist for neural networks. In particular, cascade-correlation networks (47) automatically learn how many hidden nodes to use in feed-forward networks. Like NEAT and NEAT+Q, they start with simple networks with no hidden nodes. If, after training with backpropagation, the error is above some acceptable threshold, a hidden node is added, with link weights from the inputs set to maximize the correlation between the hidden node's value and the network's error. The network is then retrained, with these correlation weights held fixed, and the process repeats, with a new hidden node added at each step, until error drops below the threshold.

By contrast, the optimal brain damage approach (82) does not complexify simple networks but rather simplifies complex ones. It does so by pruning the links that are the least *salient*, where salience is defined as the magnitude of the change in error that results from a small perturbation of the link's weight.

Decision tree and cascade-correlation methods differ from those presented in this book in that they assume the existence of a fixed set of labeled training data which can be analyzed to deduce what representations will be effective. This book focuses on reinforcement learning, for which that assumption does not hold. The next section reviews work on representation-learning methods designed to meet the particular challenges of reinforcement learning.

8.1.2 Reinforcement Learning

In reinforcement learning, no human expert is available to provide examples of what action to take in certain states. Consequently, no labeled training data is available and the agent must either search for a policy that maximizes a reward signal (as in policy search methods) or learn a value function (as in dynamic programming and temporal difference methods). Learning a value function involves computing estimated labels (i.e., value estimates for states or state-action pairs) but those labels are not fixed, since they are based on other value estimates that are also in flux.

These complications mean that representation-optimizing methods for supervised learning are not directly applicable to reinforcement learning problems. In some cases, however, it may be possible to adapt those methods to reinforcement learning. For example, Rivest and Precup (120) train cascade-correlation networks as value function approximators using temporal difference methods. Since the training examples produced by temporal difference methods appear only in sequence and quickly become stale, Rivest and Precup use a novel caching system that in effect creates a hybrid value function consisting of a table and a neural network.

This approach represents a promising way to marry the representation-optimizing capacity of cascade-correlation networks and other supervised algorithms with the power of temporal difference methods. However, it has some significant shortcomings as well. First, their approach delays the exploitation of the agent's experience, since new samples are initially added only to the cache and only intermittently used to update the network. Second, the reliance on a cache is likely to be infeasible in larger domains. Since the cache merely records which states are visited and cannot generalize value estimates, it may perform poorly in high-dimensional problems like the scheduling task. Rivest and Precup evaluate their method in a Tic-Tac-Toe domain with 3^9 states. By contrast, the scheduling domain has 100^{16} states. Third, their approach evaluates representations based only on their ability to approximate the value function. It does not directly favor representations that yield good policies, as evolutionary function approximation and adaptive tile coding (with the policy criterion) do. Relying solely on the value function as a guide to selecting policies and their representations can be very risky in practice. See Section 9.3.2 for a detailed discussion of this issue.

Beyond Rivest and Precup's work, most efforts to learn representations in reinforcement learning focus on finding the right basis functions for linear value function approximators. Value functions are rarely linear with respect to the original state features supplied to the agent. However, if the right basis functions can be found, the value function can be accurately represented with a linear function approximator. The remainder of this section surveys methods that employ this approach.

Santamaria et al. (122) apply skewing functions to state-action pairs before feeding them as inputs to a function approximator. These skewing functions make the state-action spaces non-uniform and hence make it possible to give more resolution to the most critical regions. Using various skewing functions, they demonstrate improvement in the performance of temporal difference methods. However, they do not offer any automatic way of determining how a given space should be skewed. Hence, a human designer still faces the burdensome task of manually choosing a representation, though in some domains using skewing functions may facilitate this process.

Smith (130) extends the work of Santamaria et al. by introducing a method that uses self-organizing maps to automatically learn nonlinear skewing functions for the state-action spaces of reinforcement learning agents. Self-organizing maps use unsupervised learning methods to create spatially organized internal representations of the inputs they receive. Hence, the system does not use any feedback on the performance of different skewing functions to determine which one is most appropriate. Instead it relies on the heuristic assumption that more resolution should be given to regions of the space that are more frequently visited. While this heuristic is intuitive and reasonable, it does not hold in general. For example, a reinforcement learning agent designed to respond to rare emergencies may spend most of its life in safe states where its actions have little consequence and only occasionally experience crisis states where its choices are critical. Smith's heuristic would incorrectly devote most of its resolution to representing the value function of the unimportant but frequently visited states. This issue distinguishes Smith's approach from the methods presented in this book. Evolutionary function approximation, introduced in Chapters 4 and 5, avoids this problem because it evaluates competing representations by testing them in the actual task. It explicitly favors those representations that result in higher performance, regardless of whether they obey a given heuristic. Similarly, adaptive tile coding devotes more resolution to regions where the value function or policy changes the most, independent of how frequently those regions are visited.

Mahadevan (87) advocates learning *proto-value functions*, derived from a global state space analysis. Though labeled training data is unavailable, a reinforcement learning domain nonetheless has structural properties than can be analyzed to determine effective representations. In Mahadevan's approach, the agent's experience is used to build a graph representing how states are connected in the domain. Next, a spectral analysis of the graph Laplacian is performed. The resulting eigenfunctions, or proto-value functions, are then used as basis functions for a linear function approximator trained with standard reinforcement learning methods. While the original method is applicable only to domains with discrete state spaces, a recent extension handles continuous domains as well (89; 88). In similar work, Parr et al. (109) also propose methods for automatically finding basis functions for linear function approximation, in this case by using Bellman error to automatically selecting orthogonal basis functions.

The main drawback of the proto-value function approach is that it assumes the agent has access to state transitions gathered on a random walk of the domain. These transitions are necessary to build the state graph, but may not be feasible to obtain in large or high-dimensional state spaces or domains where exploration is expensive. This problem distinguishes Mahadevan's approach from the adaptive methods presented in this book. While both evolutionary function approximation and adaptive tile coding seek the best representation for each stage in the learning process, proto-value functions are used to find the best final representation, once the structure of the domain is known.

Munos and Moore (102) present an approach to learning variable resolution function approximators based on kd-trees. Their approach is similar to adaptive tile coding, in that they repeatedly subdivide the state space into smaller and smaller regions. The primary difference is the use of piecewise-linear representations instead of tile coding. As a result, computing $V(s)$ once the right tile is located takes order of $k \ln k$ time instead of constant time. They propose a splitting rule that is similar to the value criterion used in adaptive tile coding. They also propose examining the policy to determine where to split, though their approach, unlike the policy criterion used in adaptive tile coding, does not reason about sub-tile weights and works well only in conjunction with a criterion based on the value function. In addition, their method does not reason about *when* to split tiles but instead runs dynamic programming to convergence between each split, which may be computationally inefficient. Their empirical evaluations measure final performance at each resolution but do not consider, as we do in Chapter 7, the speed of learning as measured in number of updates.

G-learning (35) also uses a tree structure to grow a value function representation. Like adaptive tile coding, it starts with a coarse representation and refines it during learning by adding new partitions. Partitions are made based on the expected change to the value function, similar to the value criterion used in adaptive tile coding. However, the method does not select splits that maximize improvement to the policy, as the policy criterion does.

Utile Suffix Memory (90) extends G-learning to automatically learn history-based representations. The tree-based representation contains a history of recent relevant observations. Statistical tests are used to determine whether a given observation is worth remembering, based on its capacity to distinguish among states with different values. Unlike the methods presented in this book and the other methods reviewed in this section, Utile Suffix Memory focuses on the problem of partial observability. In other words, the agent assumes that its state is not Markovian and that some different states will yield the same immediate observation. Consequently, the agent must remember some of its previous observations in order to disambiguate its current state. Utile Suffix Memory strives to deduce which observations to remember. However, in so doing, it also allows generalization across states with similar values, and hence takes a similar approach to adaptive tile coding.

Sherstov and Stone (126) present a tile-coding method with fixed tile sizes but variable generalization. They use the Bellman error generated by temporal difference updates to assess the reliability of the function approximator in a given region of the state or action space. This metric is used to automatically adjust the breadth of generalization for a tile-coding function approximator. An advantage of this approach is that feedback arrives immediately, since Bellman error can be computed after each update. A disadvantage is that the function approximator's representation is not selected based on its actual performance, which may correlate poorly with Bellman error.

Chow and Tsitsiklis (36) show how to compute the tile width of a uniform tiling necessary to learn an approximately optimal policy, though they make strong assumptions (e.g., that the transition probabilities are Lipschitz continuous). Like adaptive tile coding, they advocate beginning with coarse representations and refining them over time, though refinements always occur across the entire state space, such that all regions always have the same size tiles.

Like adaptive tile coding, the Parti-game algorithm (98) repeatedly partitions the state space to grow a representation suitable to the given task. However, this method is not designed to tackle reinforcement learning tasks in general. On the contrary, it applies only to tasks that consist of navigating some space to reach a goal region whose location is known to the agent *a priori*. In addition, Parti-game assumes the agent has access to a greedy local controller which allows it to travel from one tile to another. Given these assumptions, standard shortest-path graph algorithms are used to plan a path to the goal, with each step in the path executed by the greedy local controller.

8.1.3 Evolutionary Computation

Evolutionary methods, like other stochastic optimization techniques, search a space of candidate solutions for one that maximizes some fitness function. Many such methods evolve only the solution's weights and require a human designer to specify the solution's representation. However, other methods can evolve the solution's representation as well. Unlike the representation-learning supervised methods described in Section 8.1.1, this approach does not require a set of labeled data to analyze. Instead, the space of candidate representations is searched, using the given fitness function as a guide. Though many types of representations have been evolved, this section focuses on the evolution of neural networks, called *neuroevolution* (172), as it is most related to this book.

Many neuroevolutionary methods, such as Symbiotic, Adaptive Neuro-Evolution (SANE) (100) and Enforced Sub-Populations (ESP) (54), assume a fixed topology and evolve only link weights. Neuroevolutionary methods that evolve network topologies too are sometimes called *Topology and Weight Evolving Neural Networks* (TWEANNs) (133).

Perhaps the simplest of these is the Structured Genetic Algorithm (sGA) (40), in which one bit string represents each network's connection matrix and another bit string represents the weights of each link. These bit strings are then evolved using standard genetic algorithms. Hence, sGA can automatically discover which links are useful for the given task, at the same time that it evolves weights for those links. However, the number of nodes in the network is not evolved but must be set manually before evolution begins. Furthermore, the encoding scheme is not concise, since much of the genome is wasted when networks are not fully connected. More importantly, since the initial population consists of random bit strings, sGA does not complexify. Instead of bootstrapping off solutions in lower-dimensional spaces the way NEAT does, sGA must search the entire space of representations from scratch. This difficulty is exacerbated by the fact that many genomes correspond to infeasible networks, which lack even a single path from inputs to outputs.

To obtain a more concise representation than sGA, many approaches use graph-based encoding, where each network's topology and weights are captured in a variable-length genome that enumerates the network's nodes and describes their connectivity. However, performing crossover on populations with heterogeneous structure is notoriously problematic. Even if two parents have similar behavior and performance, they may represent their solutions very differently, an issue known as the *competing conventions problem* (167) and consequently crossover may have catastrophic consequences.

Due to this difficulty, some representation-learning neuroevolutionary methods simply omit crossover altogether and rely solely on mutation operators to search the space of possible solutions. For example, GeNeralized Acquisition of Recurrent Links (GNARL) (8) uses a graph-based encoding, with structural and weight mutations as the only genetic operators. Unlike in NEAT, new nodes are added without connecting then to the rest of the network. Separate mutations are required to add new links connecting these additional nodes.

Other neuroevolutionary methods preserve crossover and try to ensure that it occurs in a constructive way. For example, Pujol and Poli (113) present an approach based on Parallel Distributed Genetic Programming (PDGP) (111) which uses a graph-based encoding to evolve neural network topologies. As in genetic programming (74), entire subgraphs are swapped during crossover. The motivation for this approach is the intuition that subgraphs represent important functional units. Hence, preserving them reduces the chance that crossover will have catastrophic effects on the offspring's fitness.

Cellular encoding (57) is a neuroevolutionary method that takes a radically different approach to learning representations. Instead of encoding each neural network as a bit string or graph, it uses an *indirect encoding*. Unlike direct encodings, which explicitly list each node and link in the network, indirect encodings merely specify rules by which the network can be constructed. In the case of cellular encoding, these rules are written in a graph transformational language called a grammar tree. The transformations specified in the tree indicate how to grow the network via a developmental process akin to organic cell division. An important advantage of cellular encoding is that its genomes are very concise, since each transformation can be reused many times during the construction of a network. Concise genomes result in smaller spaces for evolution to search and therefore potentially better performance. However, catastrophic crossover remains a problem, exacerbated by the inscrutability of the genomes. Since the networks are not represented explicitly, it is difficult to analyze their structure to identify subgraphs or other features that might facilitate smarter crossover. Empirical results have shown that NEAT can dramatically outperform cellular encoding (137).

The NEAT method, overviewed in Section 2.3, is similar in some ways to other neuroevolutionary methods. Like Pujol and Poli's method, it uses a graph-based encoding. Like GNARL, it complexifies, starting with simple networks and adding new structure via mutations. It is unique, however in its approach to crossover, which relies on the notion of historical markings to identify which nodes and links correspond between two parents. It is further distinguished by its reliance on speciation to protect innovation by giving evolution a chance to optimize new structure before subjecting it to full selective pressure. Most importantly, NEAT stands out because of its impressive empirical record tackling challenging optimization tasks such as non-Markovian double pole balancing (137), game playing (139), robot control (138; 150), and data filtering in high energy physics (1; 166).

All the methods described in this section are general purpose optimization techniques. So long as a fitness function is supplied which can evaluate the quality of a given neural network, these methods can evolve networks that strive to maximize that fitness function. Hence, they are applicable to reinforcement learning tasks via the policy search approach outlined in Section 2.3. Moriarty et al. (101) provide a detailed survey of applications of evolutionary methods to reinforcement learning. Evolutionary function approximation differs from these approaches in that it strives to evolve value functions instead of policies and hence to synergistically combine evolution and learning. Evolution and learning have been combined before (as the next section details) but never, to our knowledge, to aid the discovery of good temporal difference function approximators.

8.2 Combining Evolution and Learning

This section reviews the substantial body of research that focuses on combinations of evolution and learning and discusses its relationship to evolutionary function approximation (Chapters 4 and 5). Perhaps the earliest of these is the work of Hinton and Nowlan (61), who demonstrate empirically that the Baldwin Effect can speed evolution. To do so, they devise an artificial scenario in which neural networks with a fixed number of binary connections receive positive fitness only if all their connection weights match an arbitrary pattern. For each connection, the genome can either specify the corresponding weight or leave it open to learning. Learning occurs by randomly altering unspecified weights to search for the combination that yields positive fitness. Though their approach is very simple and does not tackle reinforcement learning problems or evolve representations, it is sufficient to demonstrate the potential benefits of combining evolution and learning.

Since Hinton and Nowlan's work, many other researchers have investigated such combinations, in an effort to better understand the underlying population dynamics. For example, French and Messinger (50) present experiments that further verify the Baldwin Effect's ability to speed evolution. Their work differs from Hinton and Nowlan in that they study an artificial life domain wherein individuals control their own reproduction. In addition, in their experiments, not all traits are equally difficult to learn but rather vary over a range. Furthermore, their agents do not actually learn; instead the effects of learning are merely simulated in order to study the Baldwin Effect. Similarly, Arita and Suzuki (9) extend results demonstrating the benefits of the Baldwin Effect to non-stationary multi-agent domains. Their work focuses on the iterated prisoner's dilemma, where each agent not only evolves but can learn in response to the behavior of other agents in the population, against which it is competing.

8.2.1 Applications to Supervised Learning

Most combinations of evolution and learning aim not to reveal evolution's inner workings but rather to improve its performance on challenging problems. Much of this work focuses on supervised learning tasks, for which evolution can be combined with any supervised learning technique in a straightforward manner. For example, Boers et al. (25) introduce a method that evolves neural networks, each of which is trained with a learning method based on backpropagation. Like NEAT+Q, their method can automatically discover network topologies, though they are not evolved. On the contrary, only the learning component can alter network topologies. It does so by adding new nodes to those modules in the network deemed most "computationally deficient". The computational deficiency of a module is defined as the magnitude of the weight changes that still occur (due to backpropagation) even after substantial training. Weight changes in each network are not written back to their genomes. Hence, their system is Darwinian and exploits the Baldwin Effect. While the Boers et al. method can automate the design of neural networks, it is applicable only to supervised learning problems.

Giraud-Carrier (51) also combines evolution and learning for supervised tasks. His system, called GA-RBF, evolves *radial basis functions* (RBFs) (30). Like NEAT+Q and the Boers et al. method, GA-RBF strives to automatically find good representations. In this case, however, the role of determining the right representation is shared by evolution and learning. Each genome specifies the number and initial position of each RBF's centroids. In the learning phase, the position of each centroid is adjusted using an unsupervised clustering method and then the weights of the resulting representation are learned in a supervised fashion. GA-RBF can be implemented in either a Darwinian or Lamarckian way. However, since the RBF weights are not encoded in the genome, only the clustering phase of the learning process can be preserved across generations. The weights must necessarily be relearned each generation, in a Darwinian fashion. Because both evolution and learning are involved in determining the representation, GA-RBF is an intriguing approach. However, like the Boers et al. method, its use is restricted to supervised learning problems.

Evolino (124) is a method that combines evolution of recurrent neural networks (in which previously experienced outputs are fed back into the network) with learning on a linear output layer. Evolution occurs using Enforced Sub-Populations (ESP) (54), which co-evolves populations of neurons that are combined to formed complete networks. The weights of the linear output layer are learned via linear regression or quadratic programming. ESP is used to evolve Long-Short Term Memory networks which are heavily recurrent and designed to tackle tasks that require significant memory. As a result, Evolino excels at time series prediction and other sequential learning tasks. However, unlike NEAT+Q, it does not evolve representations.

Gruau and Whitley (56) present a combination of evolution and learning that tackles supervised learning problems but does not use supervised learning methods. Instead, it extends cellular encoding to incorporate unsupervised Hebbian learning methods (60) that adjust network weights. Though this learning method does not directly minimize network error in the supervised task, plasticity in the weights nonetheless enables evolutionary speedup via the Baldwin effect. The addition of learning to cellular encoding also creates a platform for comparing the performance of Darwinian and Lamarckian evolution, as Gruau and Whitley do across multiple supervised tasks. They find that the Lamarckian approach performs consistently better. These results are consistent with those presented in Section 4.2.4 for the mountain car domain, though they clash with those for the server job scheduling domain. However, as demonstrated in Section 4.2.5, the poor performance of Lamarckian evolution in the scheduling task stems from the instability of the networks under continual learning. This instability is much less likely to occur in supervised tasks with fixed targets, like those studied by Gruau and Whitley.

8.2.2 Applications to Reinforcement Learning

Combining evolution and learning is less straightforward in reinforcement learning, since such tasks do not provide the target values required by supervised learning. Stanley et al. (134) circumvent this problem by using unsupervised learning. Like Gruau and Whitley, they combine neuroevolution (a variation of NEAT, in this case) with Hebbian update rules. The approach is tested in a simple robot control task where the agent must remember early stimuli in order to excel. Since Hebbian updates depend on previous stimuli, they serve as a type of memory. Hence, this approach is an alternative to recurrent neural networks.

Other research focuses on ways to combine supervised learning with evolution in a way that is applicable to reinforcement learning tasks. The main difficulty is determining what to use as target values for learning. One approach to this problem is to train each member of the population to behave like its parents. McQuesten and Miikkulainen (93) present a neuroevolutionary technique based on this idea. Before its fitness evaluation, each member of the population is trained, using backpropagation, such that its outputs more closely match those of its parents on randomly selected inputs. While McQuesten and Miikkulainen's method does not evolve representations, it does provide a way to induce the Baldwin Effect in reinforcement learning tasks. Like NEAT+Q, this approach is prone to overtraining, though for different reasons. In NEAT+Q, TD updates can cause instability if networks are trained too long, as discussed in Section 4.2.5. In cultural evolution, too much training will turn offspring into copies of their parents, thus hindering evolutionary progress.

Another approach is to define a secondary supervised task that bears some relationship to the primary reinforcement learning task. Since the secondary task is supervised, target values are available for learning. Due to the relationship between the two tasks, such learning can improve performance on the primary task. Nolfi et al. (106) present a neuroevolutionary system that uses this approach. Their method adds extra outputs to the network that are designed to predict what inputs will be presented next. When those inputs actually arrive, they serve as targets for backpropagation, which adjusts the network's weights starting from the added outputs. This technique allows a network to be adjusted during its lifetime using supervised methods but relies on the assumption that forcing it to learn to predict future inputs will help it select appropriate values for the remaining outputs, which actually control the agent's behavior. Another significant restriction is that the weights connecting hidden nodes to the action outputs cannot be adjusted at all during each fitness evaluation.

Yet another strategy is to evolve self-teaching agents, which can generate their own target values for supervised learning. For example, Nolfi and Parisi (107) evolve neural networks with two sets of outputs. The first set directly controls the agent's actions. The second set produces target values which are used to adjust, via backpropagation, the weights that connect the inputs to the action outputs. Though their approach does not evolve representations, it can induce the Baldwin Effect and also create agents that cope better with non-stationary environments. Nolfi and Parisi's approach differs from evolutionary function approximation in that the former requires the agent to devise its own learning scenario, while the latter provides one based on temporal difference methods.

Ackley and Littman (2) investigate a similar approach. Using an artificial life foraging scenario, they evolve a population of "action networks" that control agents inhabiting the environment. The weights of the network are updated during each individual's lifetime using a reinforcement learning algorithm called complementary reinforcement backpropagation (CRBP) (3), an alternative to temporal difference methods. The reward signal used by CRBP is derived from "evaluation networks" that are simultaneously evolved. Like Nolfi and Parisi's work, evolution controls the learning process, though in this case learning is not supervised. Because it combines evolution with reinforcement learning, Ackley and Littman's approach is similar to evolutionary function approximation. However, the neuroevolution technique they employ does not optimize network topologies and CRBP does not learn a value function.

Like Ackley and Littman, Sasaki and Tokoro (123) combine neuroevolution with reinforcement learning. In the scenario they investigate, agents must choose whether to eat the material they encounter, which may be food or poison. The agent's actions affect the reward they receive but not their future state (i.e., what materials they encounter next). Hence, temporal difference methods are not necessary and a simplified reinforcement learning rule is used in its place. Sasaki and Tokoro also compare the performance of Darwinian and Lamarckian implementations of their system and find that Darwinian systems perform better. These results differ from those of Gruau and Whitley (56) but the difference is not surprising, since Sasaki and Tokoro's experiments focus on non-stationary environments. When the environment is in flux, the learning done by older generations may become obsolete. Hence, Darwinian systems, which start learning anew each generation, can adapt more rapidly to such changes. Sasaki and Tokoro's system is similar to evolutionary function approximation because it combines evolutionary methods with reinforcement learning. However, it does not evolve representations and, since it does not learn value functions, cannot master domains with delayed reward.

More closely related to evolutionary function approximation is *reinforced genetic programming* (44), which combines genetic programming with reinforcement learning. Unlike the work of Ackley and Littman or Sasaki and Tokoro, this system uses temporal difference methods to implement individual learning. Like evolutionary function approximation, reinforced genetic programming can be implemented in a Darwinian or Lamarckian fashion. The primary difference is the representation. Like other genetic programming methods, reinforced genetic programming relies on a tree-based representation. Each leaf of the tree corresponds to a region of the state space and has associated with it an estimate of the value function for that region. The advantage of this approach is that it harnesses existing genetic programming techniques. The disadvantage is that each weight corresponds to another parameter that must be optimized, forcing evolution to search a very high-dimensional space. Since the state space is divided into regions, the representation bears some similarity to adaptive tile coding, though the reliance on evolution to optimize that representation makes it more similar to evolutionary function approximation.

However, Downing's approach does not employ exploratory mechanisms when selecting individuals for evaluation, and hence does not optimize on-line performance.

A different approach to combining evolution and learning is *learning classifier systems* (LCS) (81). LCS methods evolve a population of rules for approximating some function. In "Pittsburgh-style" classifiers (131), each member of the population represents a candidate solution for the entire problem, i.e., an approximation of the entire function. This approach is most analogous to that used throughout this book, where each network in the population represents an entire policy (NEAT) or an entire value function (NEAT+Q). More common, however, are "Michigan-style" classifiers, for which the entire population represents one approximation of the function. In this case, each member of the population (a rule), specifies the subset of inputs for which it is applicable and approximates the function only for that subset. LCS methods are often used to tackle supervised learning problems or control problems without delayed reward. However, it can be applied to reinforcement learning tasks as well, particularly using XCS (32), a version of LCS which uses updates based on temporal difference methods. More closely related to the work in this book is NCS (31), a type of LCS which, like NEAT+Q uses neural networks. However, these methods do not evolve representations as evolutionary function approximation does.

Also related is the work of Lanzi et al. (80), which combines XCS with tile-coding: evolution optimizes the parameters of a population of tile-codings function approximators, each of which covers a different region of the state space. The use of evolution to optimize representations is similar to evolutionary function approximation. However, representing the function with an entire population and restricting each member of the population to a portion of the state space makes the method of Lanzi et al., like other Michigan-style classifiers, fundamentally distinct. The aim of automatically designing each tile coding makes this approach similar to adaptive tile coding, though it relies on evolution to do so. By contrast, adaptive tile coding demonstrates that tile coding representations can be optimized without expensive search.

Another important related method is VAPS (14). While it does not use evolutionary computation, it does combine TD methods with policy search methods. It provides a unified approach to reinforcement learning that uses gradient descent to try to simultaneously maximize reward and minimize error on Bellman residuals. A single parameter determines the relative weight of these goals. Because it integrates policy search and TD methods, VAPS is in much the same spirit as evolutionary function approximation. However, the resulting methods are quite different. While VAPS provides several impressive convergence guarantees, it does not address the question of how to represent the value function.

Other researchers have also sought to combine TD and policy search methods. For example, Sutton et al. (145) use policy gradient methods to search policy space but rely on TD methods to obtain an unbiased estimate of the gradient. Similarly, in actor-critic methods (73), the actor optimizes a parameterized policy by following a gradient informed by the critic's estimate of the value function. Like VAPS, these methods do not learn a representation for the value function.

8.3 Balancing Exploration and Exploitation

The difficulty of balancing exploration and exploitation is one of the most thoroughly studied problems in artificial intelligence. This section overviews methods for tackling this problem in k-armed bandit problems, associative search, and reinforcement learning. It also discusses their relationship to on-line evolutionary function approximation (Chapter 3).

8.3.1 *k*-Armed Bandit Problem

The simplest formulation of the exploration/exploitation problem is the *k-armed bandit* problem (153; 19; 12), in which an agent must repeatedly choose which of k arms of a slot machine, or "bandit", to pull. After each pull, the agent receives some reward, drawn from a probability distribution specific to that arm. Its goal is to maximize the total reward it receives. To do so, it must balance exploration (pulling different arms to learn more about their expected rewards) with exploitation (pulling the *greedy* arm, i.e., the one with the highest estimate of expected reward).

The k-armed bandit problem is closely related to the reinforcement learning problem. In fact, it can be described as a reinforcement learning problem in which the MDP contains only one state and each arm corresponds to an action, each of which returns the agent to that state with probability one. Hence, the k-armed bandit problem is of great interest to the reinforcement learning community and many of the approaches used to tackle it form the basis for exploratory mechanisms in reinforcement learning.

Most of these approaches are *action-value* methods (152), in which the agent maintains a running estimate of the expected reward for each arm. This estimate can be computed by simply averaging the rewards the agent has received on each previous pull of the given arm. Rather than recomputing this average after each pull, a more computationally efficient approach is to update the average incrementally. In non-stationary domains, the true expected reward can change over time, rendering older data stale. In such cases, incremental updates can use a fixed step-size parameter, causing the weight of older data to decay exponentially (22).

The simplest action-value method is ε-greedy selection (158), described in Section 3.1, in which the agent pulls a random arm with probability ε and the greedy arm with probability $1-\varepsilon$. One shortcoming of this approach is that all non-greedy arms are equally likely to be pulled, though some may have much higher estimated reward. Softmax selection, described in Section 3.2, addresses this problem by making each arm's probability of selection a function of the current estimate of its expected reward. Neither ε-greedy nor softmax selection consider the uncertainty of the agent's estimate of each arm's expected reward. Interval estimation (66), described in Section 3.3, addresses this problem by computing confidence intervals for each estimate and always selecting the arm whose interval has the highest upper bound.

Other approaches to k-armed bandit problems beyond action-value methods include *reinforcement comparison* (146). In this approach, each time the agent receives a reward, it is compared to a *reference reward*, which is the average of all previously received rewards. This difference is used to update the agent's *preference* for that arm. Preferences are used to determine each arm's probability of selection, using a Boltzmann distribution. Another approach is *pursuit* methods (152), which maintain both preferences and action-value estimates.

All of the approaches mentioned above are heuristic in nature. However, it is possible, at least in principle, for an agent to optimally balance exploration and exploitation in the k-armed bandit problem. Using Bayes' rule (17), the agent can compute the total reward and probability of occurrence for each possible chain of events for sequences of pulls of arbitrary length (20). However, this approach assumes that the agent knows *a priori* the distribution of problem instances. In addition, it is computationally intractable, as it requires traversing a tree that grows exponentially with respect to the length of the sequence of pulls.

8.3.2 Associative Search

Associative search (16; 11), also called the *contextual bandit problem* (78; 77) is an extension of the k-armed bandit problem in which there are multiple k-armed bandit problems. At each step, the agent faces one of these problems, randomly selected. The agent also receives some additional information (equivalent to state features) that allow it to identify which bandit problem it currently faces. Simple versions of associative search are no more challenging than the original k-armed bandit problem, since the agent can simply solve each problem separately and index the solution with the corresponding state information. However, if there are many states or the state features are continuous, the agent may need to effectively generalize across related states in order to perform well.

The associative search problem represents a partial step from k-armed bandit problems to the full reinforcement learning task. The agent must reason about multiple states, but its goal is still to maximize immediate reward. It need not reason about delayed reward because its actions have no effect on it. In other words, which arm the agent pulls has no bearing on which bandit problem it faces at the next step. The opposite is true in the full reinforcement problem, where the agent's action affects the state to which it transitions. Since some states can offer the agent more reward than others, its action affects, not only its immediate reward, but its opportunities for future reward.

8.3.3 Reinforcement Learning

The simplest approach to balancing exploration and exploitation in the full reinforcement learning problem is to borrow action-value methods from the k-armed bandit problem. For example, ϵ-greedy selection, softmax selection, and interval estimation can all be applied to reinforcement learning problems by simply replacing estimates of expected immediate reward with estimates of long-term value, using

Q or V. This approach ensures that, for each state the agent experiences, it will properly explore the actions available to it. However, it does not enable the agent to *seek* states where greater exploration is needed, a complication that does not arise in the k-armed bandit problem.

Recent approaches do address this issue, however. For example, Simsek and Barto (127) present an approach wherein the agent behaves greedily with respect to its current policy for a *derived MDP*, a solution to which describes the optimal way to explore the original MDP. In addition, some model-based approaches such as prioritized sweeping (97) and model-based interval estimation (142) employ *optimistic initialization* (149) to encourage the agent to travel to states that have been visited only infrequently. Some model-based methods such as E^3 (68) and *R-max* (29) find probably approximately optimal policies given only a polynomial number of samples. Recently, similar results have been obtained with a model-free method called *delayed Q-learning* (141). As in k-armed bandit problems, Bayes-optimal strategies for exploration can be computed (143; 45; 112). However, the same problems of computational intractability persist, rendering this approach impractical even for very small problems.

All of these methods differ from on-line evolutionary computation, introduced in Chapter 3, in that they balance exploration and exploitation only at the level of individual actions. This approach makes sense for standard methods where the agent learns a single value function: each time the agent acts, it need only decide whether to act greedily with respect to that value function or whether to explore. However, in evolutionary methods, the agent has a population of policies and must reason about balancing exploration and exploitation at that level.

In his classic work on evolutionary methods, Holland (62) argues that such methods already perform such a balance. The reproduction mechanism encourages exploration, since crossover and mutation result in novel genomes, but also encourages exploitation, since each new generation is based on the fittest members of the last one. However, reproduction allows evolutionary methods to balance exploration and exploitation only *across* generations, not *within* them. Once the members of each generation have been determined, they all typically receive the same evaluation time. On-line evolutionary computation addresses this shortcoming by borrowing standard action-value methods and using them to select polices for evaluation, thus boosting the reward accrued during learning.

Because it allows members of the same population to receive different numbers of evaluations, on-line evolutionary computation is also similar to previous work about optimizing noisy fitness functions. For example, Stagge (132) introduces mechanisms for deciding which individuals need more evaluations, assuming the noise is Gaussian. Beielstein and Markon (18) use a similar approach to develop tests for determining which individuals should survive. However, this area of research has a significantly different focus, since the goal is to find the best individuals using the fewest evaluations, not to maximize the reward accrued during those evaluations.

Action-value methods like ϵ-greedy have also been combined with evolutionary methods in the context of learning classifier systems (79; 92; 170). However, such mechanisms are used to select among individual actions, not to allocate evaluations among an entire population.

8.4 Feature Selection

This section reviews previous work on feature selection and compares it to Feature Selective NEAT (Chapter 6). Feature selection (24; 58) is the process of determining which subset of available inputs should be used by a machine learning algorithm. In supervised learning, these inputs typically describe examples used for training or testing. In reinforcement learning, they typically consist of state features, describing the agent's current state in the world. Feature selection is typically distinguished from *feature construction* (48; 156). In the former, we assume a set of adequate features is available but that, due to the presence of many irrelevant or redundant features, finding a minimal subset is necessary for effective learning. In the latter, adequate features are not available *a priori* but must be constructed from a description of the task or from low-level primitives.

8.4.1 Filters

One class of feature selection methods is called *filters* (63). These methods perform feature selection as a preprocessing step to some supervised learning algorithm, "filtering" out irrelevant features. This filtering is accomplished by performing some type of statistical analysis on the training data to determine which features will be most useful to the machine learning algorithm.

One of the simplest approaches is to rank the features based on correlation criteria (159) or mutual information between them and the target function (83; 41; 155). This ranking is then used to select the top k features. However, determining the right value for k can be difficult and it is often necessary to try multiple values and compare the resulting performance. Another limitation of this approach is that does not consider dependencies between the features.

The FOCUS algorithm (5) addresses this shortcoming by considering increasingly large combinations of features. It starts by considering individual features, then looks at pairs, triples, and so on until the class of each training example is disambiguated. Koller and Sahami (72) also consider dependencies between features by employing *Markov blankets*. The Markov blanket of some feature x_i is a set of features not including x_i that render x_i unnecessary. If the Markov blanket of x_i can be found, then x_i can be removed by the feature selection algorithm. Similarly, Singh and Provan (128) filter features for a Bayesian network using information-theoretic metrics. Principal components analysis (64), a statistical technique that constructs orthogonal vectors from linear combinations of features in the original space, can also be used for feature selection in machine learning (23).

8.4.2 Wrappers

While filter methods have proven effective in supervised learning problems, they are not applicable to reinforcement learning because of the absence of labeled training data. However, another class of methods, called *wrappers* (63), can, at least in principle, be used to select features in reinforcement learning. Wrappers work by searching the space of feature subsets for one that performs well in the machine learning task. They are called "wrappers" because each candidate subset is evaluated by running the given machine learning algorithm with that subset and measuring the resulting performance. Hence, the learning algorithm is a subroutine around which the feature selector is wrapped.

The primary advantage of wrappers compared to filters is that feature subsets are directly evaluated according to the actual goal of feature selection: improving the learner's ultimate performance. Even when filters accurately identify critical features, they do not consider the particular idiosyncrasies and inductive bias of the learning method that will use those features (43). The primary disadvantage of wrappers is their computational cost. Finding the right subset is NP-hard (6) and each feature subset considered requires a completely new run of the learning algorithm, though heuristic methods have been developed to try to minimize this cost (33; 99).

In principle, wrappers could be used to select features in reinforcement learning. Just as in supervised learning, each feature subset would be evaluated by running the learning algorithm with that subset, though performance would be measured by total reward accrued, rather than classification or regression error. However, this approach is highly impractical. In supervised learning, evaluating a feature subset requires only computational time. Since labeled data is typically a much scarcer resource, wrappers can be useful even if the computational cost is high. However, in reinforcement learning, evaluating a feature subset requires not only computation time but also new samples (i.e., interactions with the real world). Since samples are usually the scarcest resource, any benefit obtained by applying wrappers in reinforcement learning is unlikely to justify its cost.

8.4.3 FS-NEAT

The FS-NEAT method, introduced in Chapter 6 does not fall cleanly into either the filter or wrapper categories. It is similar to wrapper methods in that it searches for the right feature subset and evaluates candidates based on their performance in the ultimate task. However, it is practical for reinforcement learning precisely because it does *not* wrap a feature selector around the base learning method. On the contrary, it incorporates the search for a good feature subset into the search for a good network topology and good weights, without any meta-learning.

By integrating these different aspects of the task, FS-NEAT bears some resemblance to *embedded* feature selection methods (24; 58). Embedded methods, such as decision trees (116; 117), incorporate feature selection into the base learning method. However, such methods are typically similar to filters in that they rely on statistical analysis of labeled data, though not as a preprocessing step. FS-NEAT, by contrast, does not require labeled data at all. Hence, FS-NEAT represents a unique approach to the problem of feature selection, one whose advantages are particularly well suited to reinforcement learning tasks.

Recently, other feature selection methods customized to reinforcement learning have been developed (42; 75). Like FS-NEAT, these methods are neither filters nor wrappers. Instead, they are model-based methods that learn both the structure and weights of *dynamic Bayesian networks* (DBNs) that describe the transition function of the MDP. These DBNs can then be analyzed to infer what features are most useful for representing the value function. While these approaches avoid the shortcomings of both filters and wrappers, they have not so far proven successful on large tasks such as RARS, on which FS-NEAT excels. One reason is that their computational costs scale poorly with respect to the number of features available. Furthermore, since they require a large amount of data to reliably select features, their usefulness is likely restricted to cases when features sets are transferred between related tasks (75).

Chapter 9
Conclusion

This book presents a range of new methods for automating the design of effective representations for reinforcement learning. It also presents a body of empirical evidence verifying the efficacy of these new methods. This chapter begins by summarizing the conclusions that can be drawn from this evidence. Next, it discusses some negative results obtained while developing these methods. Finally, it touches on some broader implications, comparing results across chapters from a "big picture" perspective.

9.1 Primary Conclusions

First and foremost, this book demonstrates that reinforcement learning agents can automatically discover effective representations. Both evolutionary function approximation (Chapters 4 and 5) and adaptive tile coding (Chapter 7) enable such agents to autonomously revise their own representations while they are learning, without the aid of human expertise. Empirical results in multiple domains confirm the benefit of these methods. Adaptive tile coding automatically discovers representations that match the ultimate performance of the best manually designed representations and learn nearly as quickly. Evolutionary function approximation discovers representations that perform *better* than the best manually designed representations. This enables the agent to learn an approximately optimal policy in mountain car, a notoriously difficult task for neural network function approximators. These performance improvements carry over to server job scheduling, a much larger and more challenging reinforcement learning task.

Second, this book demonstrates that policy search and temporal difference methods can be combined synergistically. Rather than having to choose between alternatives with starkly different advantages and disadvantages, evolutionary function approximation makes it possible to get the best of both worlds. This approach reaps the representation-learning benefits of evolutionary methods like NEAT while simultaneously harnessing the power of temporal difference methods, which exploit

S. Whiteson: Adaptive Representations for Reinforcement Learning, SCI 291, pp. 95–104.
springerlink.com

the specific structure of the reinforcement learning problem. Furthermore, thanks to the Baldwin Effect, powerful synergies result from combining evolution in learning, yielding a system that is more than the sum of its parts.

Third, this book demonstrates that evolutionary methods can excel at on-line reinforcement learning tasks. Though such methods are typically reserved for off-line tasks, on-line evolutionary computation (Chapter 3) demonstrates that their performance can be modified to maximize the reward accrued during learning. These modifications result from another synergy between the temporal difference and policy search communities: exploratory mechanisms, traditionally used in temporal difference methods to select individual actions, can be applied to evolutionary methods to select policies for evaluation.

Fourth, this book demonstrates that feature selection can be automated in reinforcement learning. Traditional approaches to feature selection are largely inapplicable to reinforcement learning. Filters rely on labeled training data that is available only in supervised learning. Wrappers are impractical since evaluating candidate feature subsets requires new samples, not just additional computation time. However, FS-NEAT (Chapter 6) represents a new approach to feature selection, one that is particularly suited to reinforcement learning problems. By starting with a population of highly minimal networks, FS-NEAT incrementally evolves a suitable feature set at the same time that it optimizes network topology and weights. The result is a method that performs well even in the presence of large numbers of irrelevant or redundant features.

9.2 Negative Results

The preceding chapters present methods that achieved empirical success in improving the performance of reinforcement learning agents. However, in the process of developing these methods, other approaches were investigated that ultimately did not succeed. This section briefly mentions the most significant of these negative results.

9.2.1 Combining FS-NEAT with NEAT+Q

Perhaps most surprising was the poor performance that resulted from combining FS-NEAT with NEAT+Q. Such a combination, called FS-NEAT+Q, is appealing because it could allow a reinforcement learning agent to automatically and simultaneously optimize both the input and internal representations of a neural network function approximator. Yet experiments in both RARS and server job scheduling confirm that this approach performs poorly in practice. Exactly why FS-NEAT+Q fails when both FS-NEAT and NEAT+Q succeed is difficult to deduce.

However, the answer may have something to do with the fitness landscapes of degenerate networks. In early generations of FS-NEAT, every network in the population is degenerate, lacking even the basic connectivity necessary to represent a good policy. Since all these networks will perform poorly, FS-NEAT can succeed only if those networks which perform *least poorly* guide evolution towards ones that perform well. In other words, the fitness landscape around highly fit networks must include a basin of attraction that contains such degenerate networks.

FS-NEAT's empirical performance suggests that such basins do exist for networks that represent policies. Yet, in FS-NEAT+Q, networks represent value functions instead. FS-NEAT+Q's poor performance implies that degenerate value function approximators do not guide evolution toward more fit approximators. Intuitively, this result makes sense since value functions, if updated with inadequately approximated targets, can easily become unstable and divergent. If this problem arises in all networks in early generations, then evolution has no guide with which to find better representations.

9.2.2 Feature Selection in Adaptive Tile Coding

A second negative result is the performance of adaptive tile coding as a feature selector. Just as NEAT becomes feature selective if the available inputs are not initially connected to the network, adaptive tile coding should become feature selective if initial splits are not made in each dimension. In practice, all this requires is setting the number of initial tiles n to a very low value (see Section 7.2).[1] In principle, adaptive tile-coding should perform only splits that enable improvements to the value function or policy. Hence, it should never split along dimensions corresponding to irrelevant features, effectively selecting only the most useful features.

Yet in practice the data the learner uses to determine splits is quite noisy and hence spurious splits are inevitable. Overall, most of the splits are helpful, which allows it to automatically find effective representations, as described in Section 7.4. However, when even a few irrelevant features are added to the domain, its performance worsens dramatically. Examination of the learned representations reveals that, though splits along the relevant dimensions are far more likely, enough splits occur along irrelevant dimensions to incur the curse of dimensionality. Hence, unlike FS-NEAT, its performance does not scale well when the challenges of feature selection are increased.

9.2.3 Fitness Functions Based on Bellman Error

A third negative result is the performance of NEAT+Q with fitness functions based on Bellman error. In all the experiments reported in this book, the fitness function used by NEAT+Q is the average reward per episode the agent receives when controlled by the given network. As a result, NEAT selects the networks that perform best in the task, regardless of the accuracy of their value functions. That accuracy could be directly rewarded, however, if the fitness function were the inverse of the average magnitude of the Bellman error for each update.

[1] In the experiments reported in Section 7.4, n was already set quite low, to 4, though it could be set as low as 1.

If successful, such an approach would dramatically reduce the sample complexity of NEAT+Q. Even in sample-efficient NEAT+Q (Chapter 5), each network in the population must be tested in the actual domain to measure the reward it accrues. Saved experience can be used to *train* the networks, but not to *test* them, since that experience gives no information about what rewards the agent would have received if a different policy was used. By contrast, a fitness function based on Bellman error can be computed solely from saved experience. In principle, interacting with the actual domain would be necessary only initially, to build a repository of saved experience. In practice, occasionally gathering new experience is important, to ensure that the distribution of visited states in the repository roughly matches that of the agent's current policy. Nonetheless, the number of samples required is likely to be a small fraction of that needed by a fitness function based on reward.

However, experiments in both the mountain car and server job scheduling domains showed dismal performance for NEAT+Q with a fitness function based on Bellman error. To better understand why, we compared plots of average Bellman error during evolution for the two fitness functions. In both cases, Bellman error went down over time, but always remained substantial. This comparison reveals an important shortcoming of Bellman error. If the learner's Bellman error is consistently zero, it must have an optimal policy. However, having low Bellman error does not guarantee an approximately optimal policy. Rather, it seems that only a one-way implication holds in practice: higher reward implies lower Bellman error but lower Bellman error does not imply higher reward. Similar results have been obtained in the past, e.g., the VAPS method (14) performs better using fitness functions that consider reward instead of just Bellman error.

Hence, Bellman error alone is not a reliable basis for a fitness function and the tantalizing reductions in sample complexity such a fitness function promises do not appear achievable in practice. Moreover, this negative result hints at the difficulty of relying solely on value functions to solve reinforcement learning problems, one of the broader implications of this book discussed in detail below.

9.3 Broader Implications

This section discusses some of the broader implications of the results presented in this book. By comparing results across chapters, it takes more of a "big picture" perspective.

9.3.1 Stochastic vs. Deterministic Domains

Some of the methods presented in this book can be combined effectively. For example, Section 4.2.2 shows performance gains when on-line evolution is combined with evolutionary function approximation. Other combinations do not work well, as with the case of FS-NEAT+Q mentioned above. Perhaps the most interesting infeasible combination is that between on-line evolution and sample-efficient evolutionary function approximation.

The reason these approaches cannot be combined is that they are applicable to different scenarios. On-line evolution is likely to be useful only in stochastic domains because it assumes that e, the number of episodes per generation, is larger than p, the population size. In deterministic domains, individuals can be accurately evaluated in a single episode ($e = p$) so it is not possible to use previous evaluations to better balance exploration and exploitation. In principle, the value of e could be artificially inflated to allow for more exploitative episodes, though doing so would slow evolution's progress by lengthening each generation.[2]

By contrast, sample-efficient evolutionary function approximation is designed for deterministic or nearly deterministic domains. Pre-training on saved experience is possible in stochastic domains too but is unlikely to help. The evaluations necessary for estimating the noisy fitness function will already supply sufficient experience for learning. In such cases, pre-training may even be harmful since overtraining can reduce performance, as shown in Section 5.2.

This contrast suggests that the stochasticity of a domain is a critical factor in determining with which methods to tackle it. Evolutionary methods are sometimes criticized as being slow, especially in stochastic domains. Many of its successes in reinforcement learning have been in deterministic domains, e.g. (137), and recent work demonstrates that the level of stochasticity can be a critical factor in its learning speed relative to temporal difference methods (165). On-line evolution gives new hope that the performance of such methods in highly stochastic domains can be improved. On the other hand, a deterministic domain need not be tackled with evolution alone, as the sample-efficient version of evolutionary function approximation enables temporal difference learning to play an important role even when evaluations are short.

9.3.2 The Value Function Gamble

Both dynamic programming and temporal difference methods employ a strategy centered on the notion of value functions: finding the optimal value function and deriving the optimal policy from it. In small, discrete domains, this strategy is highly effective since such methods are guaranteed to converge to the optimal value function and the corresponding greedy policy is by definition optimal. When a model is known, the advantage of learning value functions is clear: dynamic programming can find the optimal policy in polynomial time (85), whereas policy search methods take exponential time in the worst case.

However, in domains that require function approximation, the benefits of a value function are much more uncertain. Some methods, like Least Squares Policy Iteration (76), guarantee convergence but assume the function approximator is linear. Furthermore, the quality of the resulting approximation depends critically on selecting appropriate basis functions. For nonlinear function approximators like neural

[2] Only ε-greedy evolution would be practical in this scenario. Softmax evolution would waste time re-evaluating individuals known with certainty to be inferior to the current champion. Interval estimation evolution would degenerate to ε-greedy evolution with $\varepsilon = 0.0$, since each individual's variance would be zero.

networks, convergence guarantees do not exist. Even if a good value function approximation is found, the corresponding greedy policy may be arbitrarily suboptimal. In such cases, using temporal difference methods means blindly gambling that the policy derived from the function approximator will perform well.

If it were necessary to choose between temporal difference and policy search methods, this difficulty could be a strong argument in favor of policy search methods, which may be less prone to catastrophic failure in practice. Though they can get trapped in local maxima, they at least directly strive to maximize reward. However, this book demonstrates that there does not have to be a trade-off between these two approaches, since it is possible to exploit the power of value function methods while still enjoying the safety of policy search. In this sense, evolutionary function approximation is a hedge against the blind gamble of temporal difference methods: the weights of individuals are adjusted using temporal difference methods but evolution is the final arbiter and it favors good policies regardless of how well they approximate the value function.

The price of such a hedge is increased sample complexity, since each candidate solution must be evaluated in the actual domain. Eliminating such evaluations requires resorting to a fitness function that examines only the value function and thus abandoning the safety of a policy search method based on reward. The negative results mentioned in the previous section highlight the practical consequences of such an approach.

This problem is exacerbated when trying to learn a representation. Nearly all of the representation-learning methods described in Section 8.1.2 examine only the value function when making representational choices. Hence, they "double down" on the gamble of temporal difference methods. They gamble not only that improving the *weights* of the function approximator will improve the policy, but that improving the *representation* of it will do so too. This approach contrasts with evolutionary function approximation, where the search for good representations is guided by performance in the domain.

Adaptive tile coding with the policy criterion shares this philosophy. Though it does not use policy search, representational choices are made based on the expected improvement to the policy, independent of how accurate the value function is. The inferior performance of the value criterion mirrors the negative results described above for NEAT+Q with a fitness function based on Bellman error. Though the value criterion's performance is not catastrophically poor, it is significantly worse than that of the policy criterion, which does not blindly focus on the value function. Hence, the results presented in this book, both for evolutionary function approximation and for adaptive tile coding, suggest that, unless stronger assumptions (i.e., a small, discrete state space or a linear function approximator) can be made, relying on the value function alone to guide an agent's policy is a dangerous proposition indeed.

9.3.3 The Role of Search in Adaptive Representations

Evolutionary function approximation and adaptive tile coding employ starkly different strategies for discovering representations. While the former relies on optimization methods to search the space of representations, the latter analyzes properties of the current representation to infer the best refinements. This contrast arises from the inherent differences in the types of representations for which the methods are designed.

Neural networks, even when they perform well, tend to operate like "black boxes." Since they are so concise, with the entire value function or policy determined by a small number of nodes and links, generalization is not controlled in any way. Altering a single weight in the network can significantly change value estimates across the entire state space. Consequently, it is difficult even for human experts to examine a neural network and deduce why it works or to meaningfully describe the role each node or link plays in the agent's value function. Similarly, when a network does not perform well, it is hard to deduce what changes to the representation might improve its performance. Hence, the most feasible strategy for finding good representations is to search for one, testing each candidate's performance in the actual domain, as NEAT+Q does.

By contrast, tile codings tend to be much more interpretable. Since generalization is strictly controlled by tile boundaries, a weight change affects value estimates only within a well-defined local region and, conversely, each value estimate is affected by only a few weights. As a result, the effects of splitting tiles are predictable and good representations can be found without search, as adaptive tile coding demonstrates. Unsurprisingly, avoiding search can greatly speed learning. For example, in the mountain car domain, adaptive tile coding requires two orders of magnitude less time than NEAT+Q to learn a good policy.[3]

This comparison demonstrates that search is not the right tool for optimizing representations for every reinforcement problem. However, it is far from a death knell. Adaptive tile coding performs well in mountain car and puddle world but these are simple domains with low dimensionality (only two state features). Scaling the method to larger problems is not trivial, as its memory requirements grow rapidly. Higher dimensional problems may strain computational resources too. The need to maintain sub-tile weights means that the cost of each update grows linearly with respect to the number of state features. By contrast, NEAT+Q excels not only at mountain car but at server job scheduling, a task with a vastly larger state space. To date, adaptive tile coding has not been tested in the server job scheduling task. However, manually designed tile coding and radial basis function have been applied to this task without success (Matthew Taylor, personal communication).

Hence, while adaptive tile coding may be useful for an important subset of reinforcement learning problems, there are likely to be many tasks whose vast complexity can be feasibly tackled only with more concise representations. As long as such

[3] The results presented in Section 7.4 show that adaptive tile coding with the policy criterion learns a good policy after about 6×10^5 updates. By contrast, results in Section 4.2.2 show that softmax NEAT+Q learns a good policy in about 1.2×10^5 episodes, using on the order of 10^7 updates. This comparison is not completely fair since adaptive tile coding uses a model while NEAT+Q does not. However, the model may not speed learning since adaptive tile coding randomly selects states to update instead of focusing updates on states experienced via the current policy, as model-free methods naturally do.

representations remain as inscrutable as neural networks, search methods will be a powerful tool for optimizing them.

9.4 Future Work

The work presented in this book opens many avenues for additional research. This section outlines a few possibilities.

9.4.1 Non-stationarity

In *non-stationary* domains, the environment can change in ways that alter the optimal policy. Since this phenomenon occurs in many real-world scenarios, it is important to develop methods that can handle it robustly. Temporal difference methods can automatically adapt to non-stationary environments so long as they constantly retain sufficient exploration. If the agent behaves completely greedily once learning plateaus, it will not be able to adapt to environmental changes. By contrast, if it continues to explore, it will discover changes in the value of its available actions and adjust its value function and policy accordingly.

However, traditional temporal difference approaches allow the agent to dynamically adjust its value function but not the *representation* of that value function. If the environment changes in ways that alter the optimal representation, then methods that automatically learn representations may perform better. By contrast, even if they are effective at the original task, manually designed representations cannot adapt to such changes. Hence, an important direction for future work is to test evolutionary function approximation and adaptive tile coding in non-stationary domains to assess their ability, not only to discover effective representations, but to adjust them in the face of environmental changes.

9.4.2 Steady-State Evolutionary Computation

The NEAT algorithm is an example of *generational* evolutionary computation, in which an entire population is is evaluated before any new individuals are bred. Evolutionary function approximation might be improved by using a *steady-state* implementation instead (49). Steady-state systems never replace an entire population at once. Instead, the population changes incrementally after each fitness evaluation, when one of the worst individuals is removed and replaced by a new offspring whose parents are among the best. Hence, an individual that receives a high score can more rapidly affect the search, since it immediately becomes a potential parent. In a generational system, that individual cannot breed until the beginning of the following generation, which might be thousands of episodes later. Hence, steady-state systems could help evolutionary function approximation perform better in on-line and non-stationary environments by speeding the adoption of new improvements. Fortunately, a steady-state version of NEAT already exists (135) so this extension is quite feasible.

9.4.3 Model-Based Reinforcement Learning

In *model-based* reinforcement learning (148; 97; 142; 42; 75; 68; 29), the agent does not directly learn a value function from experience. Instead, it uses its experience to learn an approximate model of its environment, i.e., the transition and reward functions which define the underlying MDP. Given that model, it can compute a value function, typically via dynamic programming. A critical advantage of the model-based approach is its sample efficiency. Rather than using each sample for only one update, samples are used to improve a model. Given that model, the agent can improve its value function using only computational resources, not additional samples.

The ability to trade sample complexity for computational complexity makes model-based reinforcement learning similar to the experience replay methods described in Chapter 5 and used to make evolutionary function approximation more sample-efficient. However, model-based methods have important advantages over methods that merely store and reuse experience. They can be more concise, since experience is typically integrated into a model with a fixed number of parameters. By contrast, the space required by experience replay methods grows linearly with respect to the number of samples gathered. Furthermore, model-based methods can generalize. Rather than simply replaying old experience, the model can be used to generate wholly new samples, e.g., as in the Dyna method (148), or perform the Bellman updates required by dynamic programming.

However, current methods for learning models typically assume either a small, discrete state space and use table-based representations or allow continuous state spaces but assume deterministic transitions (10). There have been preliminary efforts to learn models for domains that are both continuous and stochastic (65), but this remains an open research area. The methods presented in this book could interact in two ways with efforts to extend model-based methods to more realistic domains.

First, learning a model requires solving a similar representation problem to that addressed in this book. Just as model-free methods require a representation for the policy $\pi : S \mapsto A$ or the value function $Q : S \times A \mapsto \Re$, model-based methods require a representation for the transition function $T : S \times A \times S \mapsto [0,1]$ and the reward function $R : S \times A \times S \mapsto \Re$. Learning a model is in some ways harder than learning a value function because learning T is not a supervised learning problem but rather one of multivariate *density estimation* (125). Nonetheless, the methods presented in this book may, with modification, be used to learn good representations for models.

Second, models can be used find adaptive representations more quickly and safely. Just like experience replay, models could be used to train candidate representations without gathering additional samples. Unlike experience replay, however, models could also be used to *evaluate* candidate representations. Saved experience gives no information about what would have happened if a different action had been chosen. By contrast, models can be used to simulate entire episodes with a given policy, allowing candidate representations to be both trained and evaluated with minimal sample complexity.

9.5 Final Remarks

This book addresses a chief limitation of current reinforcement learning methods: their reliance on human expertise to design a representation for the agent's solution. It introduces new methods that enable such agents to automatically discover effective internal representations. Such methods are an integral component in the development of reinforcement learning techniques that can perform well even in the absence of human expertise. Hence, this book takes one step towards the dream of fully autonomous learning agents and truly intelligent systems.

Appendix A
Statistical Significance

To assess the statistical significance of the results presented in Chapter 4, we performed a series of Student's t-tests on each pair of methods in each domain. For each pair, we performed a t-test after every 100,000 episodes. Tables A.1 and A.2 summarize the results of these tests for the mountain car and server job scheduling domains, respectively. In each table, the values in each cell indicate the range of episodes for which performance differences were significant with 95% confidence.

Table A.1 A summary of the statistical significance of differences in average performance between each pair of methods in mountain car. Values in each cell indicate the range of episodes for which differences were significant with 95% confidence.

Episodes (x1000)	Q-Learning	Off-Line NEAT	ϵ-Greedy NEAT	Softmax NEAT	Off-Line NEAT+Q	Softmax NEAT+Q	Lamarckian NEAT+Q
Q-Learning							
Off-Line NEAT	300 to 1000						
ϵ-Greedy NEAT	200 to 1000	200 to 1000					
Softmax NEAT	200 to 1000	200 to 1000	200 to 1000				
Off-Line NEAT+Q	200 to 1000	200 to 500	200 to 1000	200 to 1000			
Softmax NEAT+Q	100 to 1000	200 to 1000	200 to 1000	900 to 1000	200 to 1000		
Lamarckian NEAT+Q	200 to 1000	200 to 1000	200 to 1000	200 to 1000	200 to 1000	100 to 1000	

Table A.2 A summary of the statistical significance of differences in average performance between each pair of methods in server job scheduling. Values in each cell indicate the range of episodes for which differences were significant with 95% confidence.

Episodes (x1000)	Q-Learning	Off-Line NEAT	ε-Greedy NEAT	Softmax NEAT	Off-Line NEAT+Q	Softmax NEAT+Q	Lamarckian NEAT+Q
Q-Learning							
Off-Line NEAT	300 to 1000						
ε-Greedy NEAT	200 to 1000	200 to 1000					
Softmax NEAT	200 to 1000	200 to 1000	not significant throughout				
Off-Line NEAT+Q	300 to 1000	300 to 500	100 to 1000	200 to 1000			
Softmax NEAT+Q	200 to 1000	200 to 1000	400 to 1000	200 to 1000	200 to 1000		
Lamarckian NEAT+Q	300 to 1000	300 to 1000	100 to 1000	100 to 1000	700 to 1000	200 to 1000	

References

[1] Aaltonen, et al.: Measurement of the top quark mass with dilepton events selected using neuroevolution at CDF. Physical Review Letters 102(15), 2001–2008 (2009)
[2] Ackley, D., Littman, M.: Interactions between learning and evolution. Artificial Life II. SFI Studies in the Sciences of Complexity 10, 487–509 (1991)
[3] Ackley, D.H., Littman, M.S.: Generalization and scaling in reinforcement learning. In: Advances in Neural Information Processing Systems, pp. 550–557 (1990)
[4] Albus, J.S.: Brains, Behavior, and Robotics. Byte Books, Peterborough (1981)
[5] Almuallim, H., Dietterich, T.G.: Learning with many irrelevant features. In: Proceedings of the Ninth National Conference on Artificial Intelligence, vol. 2, pp. 547–552 (1991)
[6] Amaldi, E., Kann, V.: On the approximation of minimizing non-zero variables or unsatisfied relations in linear systems. Theoretical Computer Science 209, 237–260 (1998)
[7] Androutsopoulos, I., Koutsias, J., Chandrinos, K.V., Spyropoulos, C.D.: An experimental comparison of naive Bayesian and keyword-based anti-spam filtering with personal e-mail messages. In: SIGIR '00: Proceedings of the 23rd Annual International ACM SIGIR Conference on Research and Development in Information Retrieval, pp. 160–167 (2000)
[8] Angeline, P.J., Saunders, G.M., Pollack, J.B.: An evolutionary algorithm that constructs recurrent neural networks. IEEE Transactions on Neural Networks 5, 54–65 (1993)
[9] Arita, T., Suzuki, R.: Interactions between learning and evolution: The outstanding strategy generated by the Baldwin Effect. Artificial Life 7, 196–205 (2000)
[10] Atkeson, C., Moore, A., Schaal, S.: Locally weighted learning for control. Artificial Intelligence Review 11, 75–113 (1997)
[11] Auer, P.: Using confidence bounds for exploitation-exploration trade-offs. J. Mach. Learn. Res. 3, 397–422 (2003)
[12] Auer, P., Cesa-Bianchi, N., Fischer, P.: Finite-time analysis of the multiarmed bandit problem. Machine Learning 47(2), 235–256 (2002)
[13] Baird, L.: Residual algorithms: Reinforcement learning with function approximation. In: Proceedings of the Twelfth International Conference on Machine Learning, pp. 30–37. Morgan Kaufmann, San Francisco (1995)
[14] Baird, L., Moore, A.: Gradient descent for general reinforcement learning. In: Advances in Neural Information Processing Systems 11. MIT Press, Cambridge (1999)
[15] Baldwin, J.M.: A new factor in evolution. The American Naturalist 30, 441–451 (1896)
[16] Barto, A.G., Sutton, R.S., Brouwer, P.S.: Associative search network: A reinforcement learning associative memory. IEEE Transactions on Systems, Man, and Cybernetics 40, 201–211 (1981)
[17] Bayes, T.: An essay towards solving a problem in the doctrine of chances. Philosophical Transactions of the Royal Society 53, 370–418 (1763)
[18] Beielstein, T., Markon, S.: Threshold selection, hypothesis tests and DOE methods. In: 2002 Congresss on Evolutionary Computation, pp. 777–782 (2002)
[19] Bellman, R.E.: A problem in the sequential design of experiments. Sankhya 16, 221–229 (1956)
[20] Bellman, R.E.: Dynamic Programming. Princeton University Press, Princeton (1957)

[21] Bellman, R.E.: A Markov decision process. Journal of Methematical Mechanics 6, 679–684 (1957)
[22] Bertsekas, D.P., Tsitsiklis, J.N.: Neural Dynamic Programming. Athena Scientific, Belmont (1996)
[23] Blum, A., Kannan, R.: Learning an intersection of k halfspaces over a uniform distribution. In: Proceedings of the Thirty-Fourth Annual Symposium on Foundations of Computer Science, pp. 312–320 (1993)
[24] Blum, A., Langley, P.: Selection of relevant features and examples in machine learning. Artificial Intelligence 97(1-2), 245–271 (1997)
[25] Boers, E., Borst, M., Sprinkhuizen-Kuyper, I.: Evolving Artificial Neural Networks using the "Baldwin Effect". In: Proceedings of the International Conference in Artificial Neural Nets and Genetic Algorithms, Ales, France (1995)
[26] Bonnlander, B.V., Weigend, A.S.: Selecting input variables using mutual information and nonparametric density estimation. In: Proceedings of the 1994 International Symposium on Artificial Neural Networks (ISANN 1994), Tainan, Taiwan, pp. 42–50 (1994), `citeseer.nj.nec.com/bonnlander96selecting.html`
[27] Boyan, J.A., Littman, M.L.: Packet routing in dynamically changing networks: A reinforcement learning approach. In: Cowan, J.D., Tesauro, G., Alspector, J. (eds.) Advances in Neural Information Processing Systems, vol. 6, pp. 671–678. Morgan Kaufmann Publishers, Inc., San Francisco (1994), `citeseer.nj.nec.com/boyan94packet.html`
[28] Boyan, J.A., Moore, A.W.: Generalization in reinforcement learning: Safely approximating the value function. In: Advances in Neural Information Processing Systems 7 (1995)
[29] Brafman, R., Tennenholtz, M.: R-max-a general polynomial time algorithm for near-optimal reinforcement learning. The Journal of Machine Learning Research 3, 213–231 (2003)
[30] Broomhead, D.S., Lowe, D.: Multivariable functional interpolation and adaptive networks. Complex Systems, 321–355 (1988)
[31] Bull, L., Hurst, J.: A neural learning classifier system with self-adaptive constructivism. In: Cantú-Paz, E., Foster, J.A., Deb, K., Davis, L., Roy, R., O'Reilly, U.-M., Beyer, H.-G., Kendall, G., Wilson, S.W., Harman, M., Wegener, J., Dasgupta, D., Potter, M.A., Schultz, A., Dowsland, K.A., Jonoska, N., Miller, J., Standish, R.K. (eds.) GECCO 2003. LNCS, vol. 2724, pp. 11–18. Springer, Heidelberg (2003)
[32] Butz, M.V., Wilson, S.W.: An algorithmic description of XCS. Soft Computing - A Fusion of Foundations, Methodologies and Applications 6(3-4), 144–153 (2002)
[33] Caruana, R., Freitag, D.: Greedy attribute selection. In: Proceedings of the Eleventh International Conference on Machine Learning, pp. 28–36 (1994)
[34] Chan, P., Stolfo, S.: Toward scalable learning with non-uniform class and cost distributions: A case study in credit card fraud detection. In: Proceedings of the Fourth International Conference on Knowledge Discovery and Data Mining, pp. 164–168 (1998)
[35] Chapman, D., Kaelbling, L.P.: Input generalization in delayed reinforcement learning: An algorithm and performance comparisons. In: Proceedings of the International Joint Conference on Artificial Intelligence (1991)
[36] Chow, C.S., Tsitsiklis, J.N.: An optimal one-way multigrid algorithm for discrete-time stochastic control. IEEE Transactions on Automatic Control 36(8), 898–914 (1991)

[37] Coons, K.K., Robatmili, B., Taylor, M.E., Maher, B.A., McKinley, K., Burger, D.: Feature selection and policy optimization for distributed instruction placement using reinforcement learning. In: Proceedings of the Seventh International Joint Conference on Parallel Architectures and Compilation Techniques (PACT), pp. 32–42 (2008)
[38] Crites, R.H., Barto, A.G.: Elevator group control using multiple reinforcement learning agents. Machine Learning 33(2-3), 235–262 (1998), citeseer.ist.psu.edu/crites98elevator.html
[39] Dalvi, N., Domingos, P., Mausam, S.S., Verma, D.: Adverserial classification. In: Proceedings of the 2004 ACM SIGKDD International Conference on Knowledge Discovery and Data Mining, pp. 99–108 (2004)
[40] Dasgupta, D., McGregor, D.: Designing application-specific neural networks using the structured genetic algorithm. In: Proceedings of the International Conference on Combinations of Genetic Algorithms and Neural Networks, pp. 87–96 (1992)
[41] Dhillon, I., Mallela, S., Kumar, R.: A divisive information-theoretic feature clustering algorithm for text classification. Journal of Machine Learning Research 3, 1265–1287 (2003)
[42] Diuk, C., Li, L., Leffler, B.: The adaptive k-meteorologists problem and its application to structure learning and feature selection in reinforcement learning. In: Proceedings of the 26th Annual International Conference on Machine Learning. ACM, New York (2009)
[43] Doak, J.: An evaluation of feature selection methods and their application to computer security. Tech. Rep. CSE-92-18, University of California at Davis (1992)
[44] Downing, K.L.: Reinforced genetic programming. Genetic Programming and Evolvable Machines 2(3), 259–288 (2001)
[45] Duff, M.: Optimal learning: Computational procedures for Bayes-adaptive markov decision processes. Ph.D. thesis. University of Massassachusetts, Amherst, MA (2002)
[46] Ertoz, L., Lazarevic, A., Eilerston, E., Lazarevic, A., Tan, P., Dokas, P., Kumar, V., Srivastava, J.: The MINDS - Minnesota Intrustion Detection System, ch. 3. MIT Press, Cambridge (2004)
[47] Fahlman, S.E., Lebiere, C.: The cascade-correlation learning architecture. In: Advances in Neural Information Processing Systems, pp. 524–532 (1990)
[48] Fawcett, T.: Feature discovery for problem solving systems. Ph.D. thesis. University of Massassachusetts, Amherst, MA (1993)
[49] Fogarty, T.C.: An incremental genetic algorithm for real-time learning. In: Proceedings of the Sixth International Workshop on Machine Learning, pp. 416–419 (1989)
[50] French, R., Messinger, A.: Genes, phenes and the Baldwin effect: Learning and evolution in a simulated population. Artificial Life 4, 277–282 (1994)
[51] Giraud-Carrier, C.: Unifying learning with evolution through Baldwinian evolution and Lamarckism: A case study. In: Proceedings of the Symposium on Computational Intelligence and Learning (CoIL 2000), pp. 36–41 (2000)
[52] Goldberg, D.E.: Genetic Algorithms in Search, Optimization and Machine Learning. Addison-Wesley, Reading (1989)
[53] Gomez, F., Burger, D., Miikkulainen, R.: A neuroevolution method for dynamic resource allocation on a chip multiprocessor. In: Proceedings of the INNS-IEEE International Joint Conference on Neural Networks, pp. 2355–2361 (2001)
[54] Gomez, F., Miikkulainen, R.: Solving non-markovian control tasks with neuroevolution. In: Proceedings of the International Joint Conference on Artificial Intelligence, pp. 1356–1361 (1999)

[55] Gordon, G.J.: Stable function approximation in dynamic programming. In: Proceedings of the Twelfth International Conference on Machine Learning, pp. 261–268 (1995)
[56] Gruau, F., Whitley, D.: Adding learning to the cellular development of neural networks: Evolution and the Baldwin effect. Evolutionary Computation 1, 213–233 (1993)
[57] Gruau, F., Whitley, D., Pyeatt, L.: A comparison between cellular encoding and direct encoding for genetic neural networks. In: Genetic Programming 1996: Proceedings of the First Annual Conference, pp. 81–89 (1996)
[58] Guyon, I., Elisseeff, A.: An introduction to variable and feature selection. Journal of Machine Learning Research 3(Mar), 1157–1182 (2003)
[59] Harvey, P.R., Booth, D.M., Boyce, J.F.: Evolving the mapping between input neurons and multi-source imagery. In: Proceedings of the 2002 Congress on Evolutionary Computation, pp. 1878–1883 (2002)
[60] Hebb, D.O.: The Organization of Behavior. Wiley, New York (1949)
[61] Hinton, G.E., Nowlan, S.J.: How learning can guide evolution. Complex Systems 1, 495–502 (1987)
[62] Holland, J.H.: Adaptation in Natural and Artificial Systems: An Introductory Analysis with Applications to Biology, Control and Artificial Intelligence. University of Michigan Press, Ann Arbor (1975)
[63] John, G.H., Kohavi, R., Pfleger, K.: Irrelevant features and the subset selection problem. In: Proceedings of the Eleventh International Conference on Machine Learning, pp. 121–129 (1994)
[64] Jolliffe, I.T.: Principal Components Analysis. Springer, New York (1986)
[65] Jong, N., Stone, P.: Model-based function approximation in reinforcement learning. In: Proceedings of the 6th International Joint Conference on Autonomous Agents and Multiagent Systems, p. 95. ACM, New York (2007)
[66] Kaelbling, L.P.: Learning in Embedded Systems. MIT Press, Cambridge (1993)
[67] Kaelbling, L.P., Littman, M.L., Moore, A.P.: Reinforcement learning: A survey. Journal of Artificial Intelligence Research 4, 237–285 (1996)
[68] Kearns, M., Singh, S.: Near-optimal reinforcement learning in polynomial time. Machine Learning 49(2), 209–232 (2002)
[69] Kephart, J.O., Chess, D.M.: The vision of autonomic computing. Computer 36(1), 41–50 (2003)
[70] Kira, K., Rendell, L.: A practical approach to feature selection. In: Proceedings of the Tenth International Conference on Machine Learning. Morgan Kaufmann, Amherst (1992)
[71] Kohl, M., Stone, P.: Policy gradient reinforcement learning for fast quadrupedal locomotion. In: Proceedings of the IEEE International Conference on Robotics and Automation, pp. 2619–2624 (2004)
[72] Koller, D., Sahami, M.: Toward optimal feature selection. In: Proceedings of the Thirteenth International Conference on Machine Learning, pp. 284–292 (1996)
[73] Konda, V.R., Tsitsiklis, J.N.: Actor-critic algorithms. In: Advances in Neural Information Processing Systems 11, pp. 1008–1014 (1999)
[74] Koza, J.R.: Genetic Programming: On the Programming of Computers by Means of Natural Selection. MIT Press, Cambridge (1992)
[75] Kroon, M., Whiteson, S.: Automatic feature selection for model-based reinforcement learning in factored MDPs. In: ICMLA 2009: Proceedings of the Eighth International Conference on Machine Learning and Applications, pp. 324–330 (2009)

[76] Lagoudakis, M.G., Parr, R.: Least-squares policy iteration. Journal of Machine Learning Research 4(2003), 1107–1149 (2003)
[77] Langford, J., Strehl, A., Wortman, J.: Exploration scavenging. In: ICML 2008, pp. 528–535 (2008)
[78] Langford, J., Zhang, T.: The epoch-greedy algorithm for multi-armed bandits with side information. In: NIPS 2008, pp. 817–824. MIT Press, Cambridge (2008)
[79] Lanzi, P.L., Colombetti, M.: An extension to the XCS classifier system for stochastic environments. In: GECCO 1999: Proceedings of the Genetic and Evolutionary Computation Conference, pp. 353–360 (1999)
[80] Lanzi, P.L., Loiacono, D., Wilson, S.W., Goldberg, D.E.: Classifier prediction based on tile coding. In: Proceedings of the 8th Annual Conference on Genetic and Evolutionary Computation, pp. 1497–1504 (2006)
[81] Lanzi, P.L., Stolzmann, W., Wilson, S.: Learning classifier systems from foundations to applications. Springer, Heidelberg (2000)
[82] LeCun, Y., Denker, J.S., Solla, S.A.: Optimal brain damage. In: Advances in Neural Information Processing Systems, pp. 598–605 (1990)
[83] Lewis, D.D.: Feature selection and feature extraction for text categorization. In: Proceedings of Speech and Natural Language Workshop, pp. 212–217 (1992)
[84] Lin, L.J.: Self-improving reactive agents based on reinforcement learning, planning, and teaching. Machine Learning 8(3-4), 293–321 (1992)
[85] Littman, M.L., Dean, T.L., Kaelbling, L.P.: On the complexity of solving Markov decision processes. In: Proceedings of the Eleventh International Conference on Uncertainty in Artificial Intelligence, pp. 394–402 (1995)
[86] Liu, C.L., Layland, J.W.: Scheduling algorithms for multiprogramming in a hard-real-time environment. Journal of the Association of Computing Machinery 20(1), 46–61 (1973)
[87] Mahadevan, S.: Samuel meets Amarel: Automating value function approximation using global state space analysis. In: Proceedings of the Twentieth National Conference on Artificial Intelligence (2005)
[88] Mahadevan, S., Maggioni, M.: Proto-value functions: A Laplacian framework for learning representation and control in Markov decision processes. Journal of Machine Learning Research 8, 2169–2231 (2007)
[89] Mahadevan, S., Maggioni, M., Ferguson, K., Osentoski, S.: Learning representation and control in continuous markov decision processes. In: Proceedings of The Twenty-First National Conference on Artificial Intelligence (2006)
[90] McCallum, A.R.: Instance-based utile distinctions for reinforcement learning. In: Proceedings of the Twelfth International Machine Learning Conference, pp. 387–395 (1995)
[91] McGovern, A., Moss, E., Barto, A.G.: Building a block scheduler using reinforcement learning and rollouts. Machine Learning 49(2-3), 141–160 (2002)
[92] McMahon, A., Scott, D., Browne, W.N.: An autonomous explore/exploit strategy. In: GECCO 2005: Proceedings of the Genetic and Evolutionary Computation Conference Workshop Program (2005)
[93] McQuesten, P., Miikkulainen, R.: Culling and teaching in neuro-evolution. In: Proceedings of the Seventh International Conference on Genetic Algorithms, pp. 760–767 (1997)
[94] McWherter, D., Schroeder, B., Ailamaki, N., Harchol-Balter, M.: Priority mechanisms for OLTP and transactional web applications. In: Proceedings of the Twentieth International Conference on Data Engineering (2004)

[95] van Mieghem, J.A.: Dynamic scheduling with convex delay costs: The generalized $c\mu$ rule. The Annals of Applied Probability 5(3), 809–833 (1995)
[96] Mitchell, T.M.: Machine Learning. McGraw-Hill, New York (1997)
[97] Moore, A., Atkeson, C.: Prioritized sweeping: Reinforcement learning with less data and less real time. Machine Learning 13, 103–130 (1993)
[98] Moore, A.W., Atkeson, C.G.: The parti-game algorithm for variable resolution reinforcement learning in multidimensional state-spaces. Machine Learning 21(3), 199–233 (1995)
[99] Moore, A.W., Lee, M.S.: Efficient algorithms for minimizing cross validation error. In: Proceedings of the Eleventh International Conference on Machine Learning, pp. 190–198 (1994)
[100] Moriarty, D.E., Miikkulainen, R.: Efficient reinforcement learning through symbiotic evolution. Machine Learning 22(11), 11–33 (1996)
[101] Moriarty, D.E., Schultz, A.C., Grefenstette, J.J.: Evolutionary algorithms for reinforcement learning. Journal of Artificial Intelligence Research 11, 199–229 (1999)
[102] Munos, R., Moore, A.: Variable resolution discretization in optimal control. Machine Learning 49, 291–323 (2002)
[103] Narendra, P.M., Fukunaga, K.: A branch and bound algorithm for feature subset selection. IEEE Transactions on Computers 26, 917–922 (1977)
[104] Ng, A.Y., Coates, A., Diel, M., Ganapathi, V., Schulte, J., Tse, B., Berger, E., Liang, E.: Inverted autonomous helicopter flight via reinforcement learning. In: Proceedings of the International Symposium on Experimental Robotics (2004)
[105] Ng, A.Y., Jordan, M.I.: PEGASUS: A policy search method for large MDPs and POMDPs. In: Proceedings of the 16th Conference on Uncertainty in Artificial Intelligence, pp. 406–415. Morgan Kaufmann Publishers Inc., San Francisco (2000)
[106] Nolfi, S., Elman, J.L., Parisi, D.: Learning and evolution in neural networks. Adaptive Behavior 2, 5–28 (1994), `http://kant.irmkant.rm.cnr.it/econets/nolfi.learn-evo.ps.Z`
[107] Nolfi, S., Parisi, D.: Learning to adapt to changing environments in evolving neural networks. Adaptive Behavior 5(1), 75–98 (1997)
[108] Novovivova, J., Pudil, P., Kittler, J.: Floating search methods in feature selection. Pattern Recognition Letters 15, 1119–1125 (1994)
[109] Parr, R., Painter-Wakefield, C., Li, L., Littman, M.: Analyzing feature generation for value-function approximation. In: Proceedings of the 24th international conference on Machine learning, p. 744 (2007)
[110] Pereira, F.B., Costa, E.: Understanding the role of learning in the evolution of busy beaver: A comparison between the Baldwin Effect and a Lamarckian strategy. In: Proceedings of the Genetic and Evolutionary Computation Conference, GECCO 2001 (2001)
[111] Poli, R.: Parallel distributed genetic programming. In: New Ideas in Optimization, pp. 403–432. McGraw-Hill, New York (1999)
[112] Poupart, P., Vlassis, N., Hoey, J., Regan, K.: An analytic solution to discrete Bayesian reinforcement learning. In: Proceedings of the Twenty-Third International Conference on Machine Learning (2006)
[113] Pujol, J.C.F., Poli, R.: Evolving the topology and the weights of neural networks using a dual representation. Applied Intelligence 8(1), 73–84 (1998)
[114] Puterman, M.L., Shin, M.C.: Modified policy iteration algorithms for discounted Markov decision problems. Management Science 24, 1127–1137 (1978)

[115] Pyeatt, L.D., Howe, A.E.: Decision tree function approximation in reinforcement learning. In: Proceedings of the Third International Symposium on Adaptive Systems: Evolutionary Computation and Probabilistic Graphical Models, pp. 70–77 (2001)
[116] Quinlan, J.R.: Induction of decision trees. Machine Learning 1(1), 81–106 (1986)
[117] Quinlan, J.R.: C4.5: Programs for Machine Learning. Morgan Kaufmann, San Mateo (1993)
[118] Radcliffe, N.J.: Genetic set recombination and its application to neural network topology optimization. Neural Computing and Applications 1(1), 67–90 (1993)
[119] Riedmiller, M.: Neural fitted Q iteration - first experiences with a data efficient neural reinforcement learning method. In: Gama, J., Camacho, R., Brazdil, P.B., Jorge, A.M., Torgo, L. (eds.) ECML 2005. LNCS (LNAI), vol. 3720, pp. 317–328. Springer, Heidelberg (2005)
[120] Rivest, F., Precup, D.: Combining TD-learning with cascade-correlation networks. In: Proceedings of the Twentieth International Conference on Machine Learning, pp. 632–639. AAAI Press, Menlo Park (2003)
[121] Rumelhart, D.E., Hinton, G.E., Williams, R.J.: Learning internal representations by error propagation. In: Parallel Distributed Processing, pp. 318–362. MIT Press, Cambridge (1986)
[122] Santamaria, J., Sutton, R., Ram, A.: Experiments with reinforcement learning in problems with continuous state and action spaces. Adaptive Behavior 6(2) (1998), citeseer.nj.nec.com/26475.html
[123] Sasaki, T., Tokoro, M.: Evolving learnable neural networks under changing environments with various rates of inheritance of acquired characters: Comparison between Darwinian and Lamarckian evolution. Artificial Life 5(3), 203–223 (1999)
[124] Schmidhuber, J., Wierstra, D., Gomez, F.J.: Evolino: Hybrid neuroevolution / optimal linear search for sequence learning. In: Proceedings of the Nineteenth International Joint Conference on Artificial Intelligence, pp. 853–858 (2005)
[125] Scott, D.: Multivariate Density Estimation. Wiley Interscience, Hoboken (1992)
[126] Sherstov, A.A., Stone, P.: Function approximation via tile coding: Automating parameter choice. In: Proceedings of the Symposium on Abstraction, Reformulation, and Approximation, pp. 194–205 (2005)
[127] Simsek, O., Barto, A.G.: An intrinsic reward mechanism for efficient exploration. In: Proceedings of the Twenty-Third International Conference on Machine Learning (2006)
[128] Singh, M., Provan, G.M.: Efficient learning of selective bayesian network classifiers. In: Proceedings of the Thirteenth International Conference on Machine Learning, pp. 453–461 (1996)
[129] Smart, W.D., Kaelbling, L.P.: Practical reinforcement learning in continuous spaces. In: Proceedings of the Seventeeth International Conference on Machine Learning, pp. 903–910 (2000)
[130] Smith, A.J.: Applications of the self-organizing map to reinforcement learning. Journal of Neural Networks 15, 1107–1124 (2002)
[131] Smith, S.: Flexible learning of problem solving heuristics through adaptive search. In: Proceedings of the Eighth International Joint Conference on Articial Intelligence, pp. 422–425 (1983)
[132] Stagge, P.: Averaging efficiently in the presence of noise. Parallel Problem Solving from Nature 5, 188–197 (1998)

[133] Stanley, K.O.: Efficient evolution of neural networks through complexification. Ph.D. thesis, Department of Computer Sciences, The University of Texas at Austin, Austin, TX (2004)
[134] Stanley, K.O., Bryant, B.D., Miikkulainen, R.: Evolving adaptive neural networks with and without adaptive synapses. In: Proceeedings of the 2003 Congress on Evolutionary Computation (CEC 2003), vol. 4, pp. 2557–2564 (2003)
[135] Stanley, K.O., Bryant, B.D., Miikkulainen, R.: Real-time neuroevolution in the NERO video game. IEEE Transactions on Evolutionary Computation 9(6), 653–668 (2005)
[136] Stanley, K.O., Kohl, N., Sherony, R., Miikkulainen, R.: Neuroevolution of an automobile crash warning system. In: Proceedings of the Genetic and Evolutionary Computation Conference, pp. 1977–1984 (2005)
[137] Stanley, K.O., Miikkulainen, R.: Evolving neural networks through augmenting topologies. Evolutionary Computation 10(2), 99–127 (2002)
[138] Stanley, K.O., Miikkulainen, R.: Competitive coevolution through evolutionary complexification. Journal of Artificial Intelligence Research 21, 63–100 (2004)
[139] Stanley, K.O., Miikkulainen, R.: Evolving a roving eye for go. In: Proceedinngs of the Genetic and Evolutionary Computation Conference (2004)
[140] Stone, P., Sutton, R.S., Kuhlmann, G.: Reinforcement learning in robocup-soccer keepaway. Adaptive Behavior 13(3), 165–188 (2005)
[141] Strehl, A., Li, L., Wiewiora, E., Langford, J., Littman, M.: PAC model-free reinforcement learning. In: Proceedings of the 23rd international conference on Machine learning, p. 888. ACM, New York (2006)
[142] Strehl, A., Littman, M.: An empirical evaluation of interval estimation for markov decision processes. In: Proceedings of the Sixteenth IEEE International Conference on Tools with Artificial Intelligence, pp. 128–135 (2004)
[143] Strens, M.: A Bayesian framework for reinforcement learning. In: Proceedings of the Seventeenth International Conference on Machine Learning, pp. 943–950 (2000)
[144] Sutton, R.: Generalization in reinforcement learning: Successful examples using sparse coarse coding. In: Advances in Neural Information Processing Systems 8, pp. 1038–1044 (1996)
[145] Sutton, R., McAllester, D., Singh, S., Mansour, Y.: Policy gradient methods for reinforcement learning with function approximation. In: Advances in Neural Information Processing Systems, pp. 1057–1063 (2000)
[146] Sutton, R.S.: Temporal credit assignment in reinforcement learning. Ph.D. thesis, University of Massachusetts, Amherst, MA (1984)
[147] Sutton, R.S.: Learning to predict by the methods of temporal differences. Machine Learning 3, 9–44 (1988)
[148] Sutton, R.S.: Integrated architectures for learning, planning, and reacting based on approximating dynamic programming. In: Proceedings of the Seventh International Conference on Machine Learning, pp. 216–224 (1990)
[149] Sutton, R.S., Barto, A.G.: Reinforcement Learning: An Introduction. MIT Press, Cambridge (1998)
[150] Taylor, M.E., Whiteson, S., Stone, P.: Comparing evolutionary and temporal difference methods in a reinforcement learning domain. In: GECCO 2006: Proceedings of the Genetic and Evolutionary Computation Conference, pp. 1321–1328 (2006)
[151] Tesauro, G.: TD-gammon, a self-teaching backgammon program achieves master-level play. Neural Computation 6, 215–219 (1994)

[152] Thathachar, M.A.L., Sastry, P.S.: A new approach to the design of reinforcement schemes for learning automata. IEEE Transactions on Systems, Man, and Cybernetics 15, 168–175 (1995)
[153] Thompson, W.R.: On the likelihood that one unknown probability exceeds another in view of the evidence of two samples. Biometrika 25, 285–294 (1933)
[154] Timin, M.E.: The robot auto racing simulator (1995), `http://rars.sourceforge.net`
[155] Torkkola, K.: Feature extraction by non-parametric mutual information maximization. Journal of Machine Learning Research 3, 1415–1438 (2003)
[156] Utgoff, P.E.: Feature construction for game playing. In: Machines that Learn to Play Games, pp. 131–152. Nova Science Publishers, Bombay (2001)
[157] Walsh, W.E., Tesauro, G., Kephart, J.O., Das, R.: Utility functions in autonomic systems. In: Proceedings of the International Conference on Autonomic Computing, pp. 70–77 (2004)
[158] Watkins, C., Dayan, P.: Q-learning. Machine Learning 8(3-4), 9–44 (1992)
[159] Weston, J., Elisseff, A., Schoelkopf, B., Tipping, M.: Use of the zero norm with linear models and kernel methods. Journal of Machine Learning Research 3, 1439–1461 (2003)
[160] Whiteson, S., Stone, P.: Adaptive job routing and scheduling. Engineering Applications of Artificial Intelligence 17(7), 855–869 (2004)
[161] Whiteson, S., Stone, P.: Evolutionary function approximation for reinforcement learning. Journal of Machine Learning Research 7(May), 877–917 (2006)
[162] Whiteson, S., Stone, P.: On-line evolutionary computation for reinforcement learning in stochastic domains. In: GECCO 2006: Proceedings of the Genetic and Evolutionary Computation Conference, pp. 1577–1584 (2006)
[163] Whiteson, S., Stone, P.: Sample-efficient evolutionary function approximation for reinforcement learning. In: AAAI 2006: Proceedings of the Twenty-First National Conference on Artificial Intelligence, pp. 518–523 (2006)
[164] Whiteson, S., Stone, P., Stanley, K.O., Miikkulainen, R., Kohl, N.: Automatic feature selection in neuroevolution. In: GECCO 2005: Proceedings of the Genetic and Evolutionary Computation Conference, pp. 1225–1232 (2005)
[165] Whiteson, S., Taylor, M.E., Stone, P.: Critical factors in the empirical performance of temporal difference and evolutionary methods for reinforcement learning. In: Autonomous Agents and Multi-Agent Systems (2009)
[166] Whiteson, S., Whiteson, D.: Machine learning for event selection in high energy physics. Engineering Applications of Artificial Intelligence 22, 1203–1217 (2009)
[167] Whitley, D.: Genetic algorithms and neural networks. In: Genetic Algorithms in Engineering and Computer Science, pp. 203–216. Wiley, Hoboken (1995)
[168] Whitley, D., Gordon, S., Mathias, K.: Lamarckian evolution, the Baldwin effect and function optimization. In: Davidor, Y., Männer, R., Schwefel, H.-P. (eds.) PPSN 1994. LNCS, vol. 866, pp. 6–15. Springer, Heidelberg (1994)
[169] Wildstrom, J., Stone, P., Witchel, E., Mooney, R.J., Dahlin, M.: Towards self-configuring hardware for distributed computer systems. In: The Second International Conference on Autonomic Computing, pp. 241–249 (2005)
[170] Wilson, S.W.: Explore/exploit strategies in autonomy. In: From Animals to Animats 4: Proceedings of the Fourth International Conference on Simulation of Adaptive Behavior (1996)

[171] Yamasaki, K., Sekiguchi, M.: Clear explanation of different adaptive behaviors between Darwinian population and Lamarckian population in changing environment. In: Proceedings of the Fifth International Symposium on Artificial Life and Robotics, vol. 1, pp. 120–123 (2000)
[172] Yao, X.: Evolving artificial neural networks. Proceedings of the IEEE 87(9), 1423–1447 (1999)
[173] Zhang, W., Dietterich, T.G.: A reinforcement learning approach to job-shop scheduling. In: Proceedings of the 1995 Joint Conference on Artificial Intelligence, pp. 1114–1120 (1995)
[174] Zweben, M., Fox, M. (eds.): Intelligent Scheduling. Morgan Kaufmann, San Francisco (1998)

GPSR Compliance

The European Union's (EU) General Product Safety Regulation (GPSR) is a set of rules that requires consumer products to be safe and our obligations to ensure this.

If you have any concerns about our products, you can contact us on ProductSafety@springernature.com

In case Publisher is established outside the EU, the EU authorized representative is:

Springer Nature Customer Service Center GmbH
Europaplatz 3
69115 Heidelberg, Germany

Batch number: 10620141

Printed by Printforce, the Netherlands